Introduction to Ray, Wave, and Beam Optics with Applications

Online at: https://doi.org/10.1088/978-0-7503-5497-4

IOP Series in Advances in Optics, Photonics and Optoelectronics

SERIES EDITOR

 **Professor Rajpal S Sirohi** Consultant Scientist

About the Editor

Rajpal S Sirohi is currently working as a faculty member in the Department of Physics, Alabama A&M University, Huntsville, AL, USA. Prior to this, he was a consultant scientist at the Indian Institute of Science, Bangalore, and before that he was Chair Professor in the Department of Physics, Tezpur University, Assam. During 2000–2011, he was an academic administrator, being vice-chancellor to a couple of universities and the director of the Indian Institute of Technology, Delhi. He is the recipient of many international and national awards and the author of more than 400 papers. Dr Sirohi is involved with research concerning optical metrology, optical instrumentation, holography, and the speckle phenomena.

About the series

Optics, photonics, and optoelectronics are enabling technologies in many branches of science, engineering, medicine, and agriculture. These technologies have reshaped our outlook and our ways of interacting with each other, and have brought people closer together. They help us to understand many phenomena better and provide deeper insight into the functioning of nature. Further, these technologies themselves are evolving at a rapid rate. Their applications encompass very large spatial scales, from nanometers to the astronomical scale, and a very large temporal range, from picoseconds to billions of years. This series on advances in optics, photonics, and optoelectronics aims to cover topics that are of interest to both academia and industry. Some of the topics to be covered by the books in this series include biophotonics and medical imaging, devices, electromagnetics, fiber optics, information storage, instrumentation, light sources, charge-coupled devices (CCDs) and complementary metal oxide semiconductor (CMOS) imagers, metamaterials, optical metrology, optical networks, photovoltaics, free-form optics and its evaluation, singular optics, cryptography, and sensors.

About IOP ebooks

The authors are encouraged to take advantage of the features made possible by electronic publication to enhance the reader experience through the use of color, animation, and video and by incorporating supplementary files in their work.

A full list of titles published in this series can be found here: https://iopscience.iop.org/bookListInfo/series-on-advances-in-optics-photonics-and-optoelectronics.

Introduction to Ray, Wave, and Beam Optics with Applications

Shanti Bhattacharya

*Department of Electrical Engineering, Indian Institute of Technology Madras,
Chennai 600 036, Tamil Nadu, India*

IOP Publishing, Bristol, UK

ISBN   978-0-7503-5497-4 (ebook)
ISBN   978-0-7503-5495-0 (print)
ISBN   978-0-7503-5498-1 (myPrint)
ISBN   978-0-7503-5496-7 (mobi)

DOI   10.1088/978-0-7503-5497-4

Version: 20241101

IOP ebooks

British Library Cataloguing-in-Publication Data: A catalogue record for this book is available from the British Library.

Published by IOP Publishing, wholly owned by The Institute of Physics, London

IOP Publishing, No.2 The Distillery, Glassfields, Avon Street, Bristol, BS2 0GR, UK

US Office: IOP Publishing, Inc., 190 North Independence Mall West, Suite 601, Philadelphia, PA 19106, USA

To my students.

Contents

Preface

This book is an outcome of the course on optical engineering that I have been teaching since 2010 at IIT Madras. Every year, different examples, ideas and perspectives have been added to my optical engineering class notes. When I received a request from IOP to write a textbook in 2022, it seemed obvious that this would be the topic that I would write it on. There was one more reason to write this book. Whilst teaching the course, I had access to many wonderful books but did not find a single textbook that seemed to cover all the points that I believe a student being introduced to this topic ought to know. It is my hope this book will act as that resource, providing a solid starting point in basic optical engineering from which students can springboard into whichever direction or depth they need to.

In addition, more and more scientists and engineers from a wide array of fields find the need to use optics, design optical experiments or work with an optical engineer to help solve their problems or meet their product goals. This book will enable them to either tackle the optics themselves or have more effective conversations with the optical engineers they are working with.

This book will present the basics of ray and wave optics. It will cover these topics in two parts. The first part (corresponding to the first four chapters) will look at geometric optics, that is treating light as rays and presenting ways of designing conventional optical systems (made with mirrors and refractive lenses). The book will provide exercises that can be done along with the content using an optical design software such as Zemax or OSLO. Using the design software will make learning easier, as the different concepts can be modelled and explained through the graphs that will be generated by the software. This part will provide readers with the basics to start designing and analysing actual optical systems. The second part of the book (chapters 5–9) changes focus and starts looking at light when its wave nature dominates. In this part concepts such as interference and diffraction will be discussed. Different types of beams will be introduced very briefly. The Gaussian beam will be dealt with in some detail but even here, rather than providing detailed theoretical derivations, an intuitive understanding of the beam will be presented. The idea is to present these beams in such a way that their properties are understood, rather than give a mathematical justification for their existence. I will also introduce ways in which such beams can be created. This knowledge, along with the beam properties, will give engineers the ability to use these beams in various applications.

Acknowledgements

It is mostly with pleasure and sometimes with surprise that I note each rendition of the course that I teach on optical engineering unveils something new or unknown (to me) about this fascinating subject. This realisation would often hit in the middle of a class, after a seemingly innocuous question. I do not believe that my experience is going to be any different in the future as well. I am grateful for the curiosity and questions of the students who have taken the course over the years, which have made teaching such an exciting experience. I also have to thank my research students who constantly push the boundaries of what I think I know.

Images play a crucial role in textbooks by clarifying concepts, especially difficult ones. Most of the pictures of this book were generated by my PhD student Bagath Chandraprasad. It was a pleasure watching my roughly drawn sketches and basic plots be transformed into pieces of art. I cannot thank him enough for the excellent images which greatly enhance the quality of the book. Since Bagath is credited with creating almost all the figures and pictures, only those that have been created by others have been specifically credited.

The pictures and photographs credited to SB were generated or created by me. In addition, I thank my other students, namely Jerin G George (JG), for help with some of the pictures in chapter 8 and for his inputs that helped with the formulation of section 7.6, Mathu Mathi (MM) for some figures in chapter 2, Susan Thomas (ST) for generating the Zemax distortion plots and the data for studying the effect of lens shape on aberrations in chapter 4 and Naveen K Pothapakula (NP) for generating the data that were used to create the optical coherence technology (OCT) related-graphs in chapter 9. I also thank my former PhD student Raghu Dharmavarapu (RD) for the figure demonstrating astigmatism in chapter 4. Finally, I thank my daughter Sumitra (SumiB) who drew three of the figures, appearing in chapters 8 and 9. As usual, it was fun working on a project together.

I am grateful to Professor Sirohi and IOP for the invitation to write this book. I had been wanting to write a textbook on this topic for a few years and their timely request is what finally got me off my derrière. Or perhaps I should say got me on my derrière, as this task involved a lot of sitting down! I thank my editors: Ms Ashley Gasque who first reached out to me and was so positive about my proposal, and to Ms Bethany Hext, my current editor, for her encouragement and support.

I would like to thank my friends for just being there. Spending time with them is what kept me sane whilst writing a book, amidst a full working schedule and countless other responsibilities. In particular, I would like to thank Namita for all her sage advice on writing and keeping to a schedule. I did slip up a little on the latter but without her inputs I fear this book would have taken much longer to finish!

I would also like to place on record my gratitude to my institute IIT Madras for allowing me to take a sabbatical during the initial phase of writing. Finally, I thank my husband, who took in good humour the many times I disrupted his plans with the refrain, 'But I have to finish my book!' Thank you Subbu!

Author biography

Shanti Bhattacharya

Shanti Bhattacharya obtained her PhD in physics from the Indian Institute of Technology, Madras, in 1997. She was awarded the Alexander von Humboldt award in 1998 and worked at the Technical University of Darmstadt, Germany, for several years. She subsequently joined Analog Devices, Cambridge, USA, where she worked as a design engineer. She is currently a Professor at the Department of Electrical Engineering, IIT Madras. She has served on the board of OSA (now known as Optica) and is currently an Associate Editor of *Optical Engineering* and the *Journal of Optical Microsystems*, as well as being a member of the editorial board of the *Journal of Optics (India)*. Her current research interests are the design and fabrication of dielectric meta and diffractive optics, optical MEMS and studies relating to imaging techniques. While she loves her work with light, she also loves her breaks, which more often than not involve escaping to the Himalayas for a while.

List of abbreviations

AFoV	angular field of view
AS	aperture stop
BB	Bessel beam
BFL	back focal length
CCD	charge-coupled device
CLC	circle of least confusion
CMOS	complementary metal oxide semiconductor
DOE	diffractive optical element
DOF	depth of focus
EFL	effective focal length
EPD	entrance pupil diameter
EUT	element under test
FDOCT	Fourier domain optical coherence tomography
FFL	front focal length
FoV	field of view
FT	Fourier transform
FTS	Fourier transform spectrometry
FTIR	frustrated total internal reflection
FWHM	full-width half maximum
GB	Gaussian beam
IR	infrared
IFTA	inverse Fourier transform algorithm
LED	light-emitting diode
LG	Laguerre–Gaussian
LSA	longitudinal spherical aberration
LSCM	laser scanning confocal microscopy
LSFM	light sheet fluorescence microscopy
MTF	modulation transfer function
NA	numerical aperture
OAM	orbital angular momentum
OCT	optical coherence technology
OPD	optical path difference
OPL	optical path length
PD	photo detector
PP	principal plane
ROC	radius of curvature
SAM	spin angular momentum
SNR	signal-to-noise ratio
SLM	spatial light modulator
STED	stimulated emission depletion
TDOCT	time domain optical coherence tomography
TCOMA	transverse coma
TSA	transverse spherical aberration
WFM	widefield fluorescence microscopy

IOP Publishing

Introduction to Ray, Wave, and Beam Optics with Applications

Shanti Bhattacharya

Chapter 1

Introduction

Gone are the days where the primary purpose of light was illumination. Today light is used for imaging, as sensors or is shaped into beams with interesting and non-intuitive behaviour for a variety of applications. Optics plays an important role in many sophisticated and esoteric systems but, more importantly, it exists in systems that we use or encounter every day. In fact, its presence may not even be obvious to the casual observer. The purpose of this book is to teach scientists and engineers the language of light and optics to help them communicate and understand each other better.

What is light? This is a question that has been asked for centuries. In 1951, Albert Einstein himself made this statement:

> All the fifty years of conscious brooding have brought me no closer to the answer to the question: What are light quanta? Of course today every rascal thinks he knows the answer, but he is deluding himself.

While Einstein was talking about light quanta, it is probably safe to say that a true understanding of light still eludes us. Depending on the circumstances, light can be thought to act as particles or waves. It is described typically as an electromagnetic wave in terms of the electric field **E** with

$$\mathbf{E} = E_0 \cos(wt - kz)\hat{x}. \tag{1.1}$$

For this first part of this book, we will mostly be concerned about the behaviour of light in a regime that does not need the intricacies of electromagnetics to explain our observations. This area of geometric or ray optics can be considered an approximation to wave optics that is valid when the wavelength of light tends to zero with respect to the sizes of the objects it is interacting with. Of course, the postulates described below can be explained using the wave nature of light. However, the important point to note is that they can be used without delving

doi:10.1088/978-0-7503-5497-4ch1

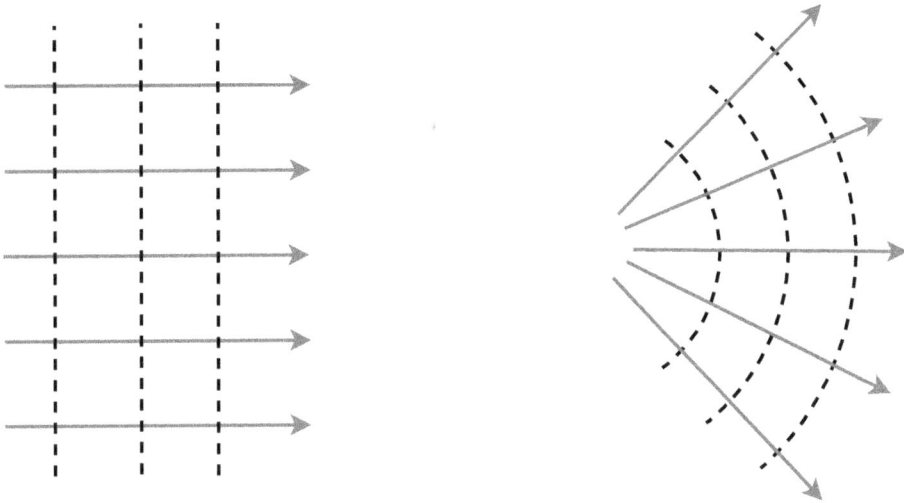

Figure 1.1. Wavefront normals (in red), indicating the rays of geometric optics. The wavefronts are shown as dashed lines. The figure on the left is a plane wave and the one on the right is a spherical wave.

into detailed theory. Visualisation of the travel of light is helped by keeping in mind that rays are the normals to a wavefront and are, therefore, very much related to the wave nature of light. Figure 1.1 shows the wavefronts of an ideal plane wave and a spherical wave, with their associated rays.

Ray optics can be used to explain simple every-day experiences such as reflection and transmission. For example, consider standing in front of a window looking out. We often can see a reflection of ourselves, as well as the scenery outside the window. Ray optics could be used to explain why the reflection is visible at a certain angle but it cannot explain how the anti-reflection coating on a pair of spectacles works or why they are tinged with a particular colour. The wave nature of light is required to explain such phenomena.

1.1 Postulates of geometric optics

Refractive index

Every dielectric medium has a parameter called the refractive index that is associated with it. The refractive index, usually denoted by n, is defined as the ratio of the velocity of light in air to the velocity of light in that medium. To understand the origin of n, one can consider a transparent dielectric medium consisting of atoms with an electron cloud around each one of them. Without going into a detailed atomic look at such a material, we can rather simply interpret the interaction of light with it by imagining that the electrons are held by springs, and it is these spring constants that determine the refractive index of the material. In a homogeneous medium, the spring constants would be identical throughout the material. When light is incident on the material, it sets the electron cloud into oscillation. If the frequency of the incident light does not match an atomic transition (which is the case for a transparent medium), the whole system can be thought of as an oscillating dipole that will re-radiate at the same frequency as the incident wave. However, there could be a phase lag between the incident and radiated field due to the 'spring constants' of the bound electrons. The constants are related to the charge and mass of an electron, the number of charges per unit volume and the resonant frequency of the bound electrons. The resulting field is nothing other than the sum of the fields generated by each oscillating dipole and is identical to a wave that travelled at a lower velocity in the medium. A detailed explanation is available in chapter 30 of [1] and also in [2]. The relevance of this apparent change in velocity in the medium is that the phase acquired by a wave travelling through a particular length of the medium compared to the same distance in air would be different. This is exploited to change the shape of a wavefront by using optical elements of varying thickness.

Light travels in straight lines
This statement is incomplete without adding the phrase 'within a medium of constant refractive index'. Clearly light would not travel in a straight line in a medium whose refractive index was a function of position $n(x, y, z)$. The way light travels through a material is determined by its properties. Alternatively, one could define its properties in terms of how light travels through the material. If a medium is homogeneous, its properties are independent of position, whereas properties of an isotropic material are independent of direction. Homogeneity and isotropy [3, 4] are of course scale dependent and these statements are true over sizes that are not sub-atomic in nature. Take, for example, figure 1.2. In the figure on the left, a homogeneous, non-isotropic pattern is drawn. Consider an observer placed somewhere within that pattern. Irrespective of where the observer is, the pattern or the environment around the observer can be considered identical, although not the same in every direction. The figure on the right, on the other hand, is a pattern that is clearly not homogeneous but is isotropic with respect to the centre. The positions of atoms in a material play an important role in all the properties (mechanical, optical, electrical), which is explored in detail in solid state physics [5].

In this book, we are interested in materials such as glass, with good optical properties, such as high transparency. Glass, ceramics and metals are examples of homogeneous materials that are also isotropic. In the case of glass, its refractive index is neither a function of position nor direction. Consider a birefringent material

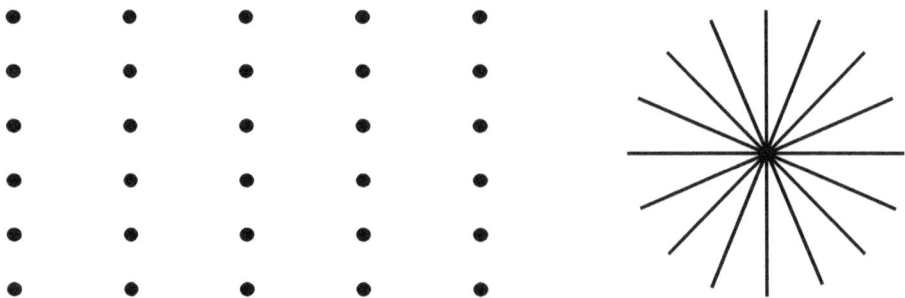

Figure 1.2. Patterns to explain the homogeneity and isotropic properties of materials.

Table 1.1. Optical material properties.

State of polarisation	Material property	What light sees
Any	Homogeneous and isotropic	n
Linear	Homogeneous and anisotropic	Different but constant n along each axes
Any	Inhomogeneous	Different n at every point

that has a different value of n along orthogonal axes of the material. The material is still considered homogeneous, as its refractive index is not a function of position but has two different constant values throughout the material. It is easier to understand these properties by considering the interaction of polarised light with the material, as shown in table 1.1. Light of any polarisation would see a single n for a homogeneous, isotropic material.

As is so often the case, fundamental electromagnetics can be used to better understand the concepts of homogeneity and isotropy in materials. The constitutive relations between the electric displacement ($\vec{D}$) and electric field ($\vec{E}$) or magnetic flux density ($\vec{B}$) and magnetic field ($\vec{H}$) are given by the equations

$$\vec{D} = \epsilon\vec{E} \tag{1.2}$$

and

$$\vec{B} = \mu\vec{H}, \tag{1.3}$$

where ϵ and μ represent the permittivity and permeability of the material, respectively.

In this form, it might appear that $\vec{D}$ is always parallel to $\vec{E}$ or $\vec{B}$ to $\vec{H}$. However, this is only true for homogeneous, isotropic materials. The general form of equation (1.2) represents ϵ as a tensor rather than a scalar, as seen in equation (1.4):

$$\begin{pmatrix} D_x \\ D_y \\ D_z \end{pmatrix} = \begin{pmatrix} \epsilon_{xx} & \epsilon_{xy} & \epsilon_{xz} \\ \epsilon_{yx} & \epsilon_{yy} & \epsilon_{yz} \\ \epsilon_{zx} & \epsilon_{zy} & \epsilon_{zz} \end{pmatrix} \begin{pmatrix} E_x \\ E_y \\ E_z \end{pmatrix}. \tag{1.4}$$

Homogeneous materials have uniform properties but different permittivity values in different directions. Therefore, the permittivity tensor reduces to

$$\begin{pmatrix} D_x \\ D_y \\ D_z \end{pmatrix} = \begin{pmatrix} \epsilon_{xx} & 0 & 0 \\ 0 & \epsilon_{yy} & 0 \\ 0 & 0 & \epsilon_{zz} \end{pmatrix} \begin{pmatrix} E_x \\ E_y \\ E_z \end{pmatrix}. \tag{1.5}$$

Isotropic materials are a special case of homogeneous materials, in which the electrical properties are both uniform and identical in all directions, leading to a scalar permittivity value, or in other words, $\epsilon_{xx} = \epsilon_{yy} = \epsilon_{zz}$.

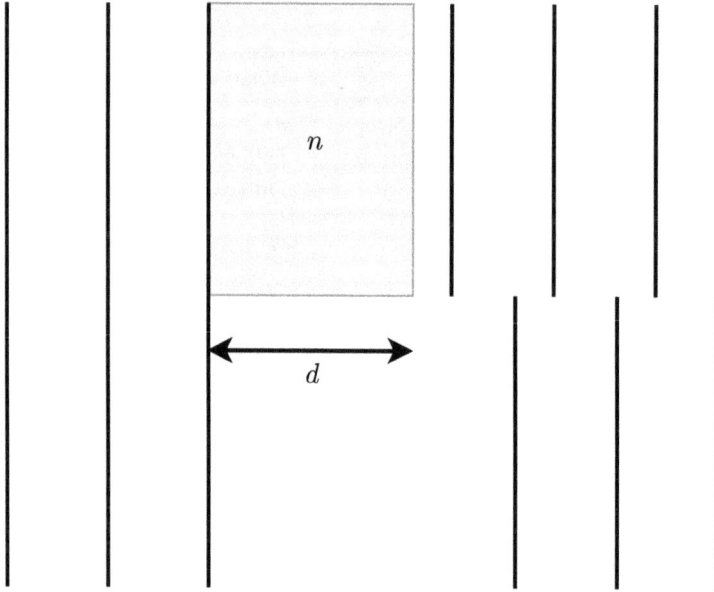

Figure 1.3. The effect of travelling through a glass slab of thickness d, and refractive index n. The light blue region represents the transparent slab. The black vertical lines are the wavefronts of light. They are only shown in the region before and after the slab.

Path length

When dealing with light, the phase acquired by the ray as it travels is important. This means that both the physical distance and the refractive index play a role. This idea is captured through the parameter optical path length (OPL). The OPL can be thought of as the equivalent distance a ray would travel in air, in the time it travels from A to B through a medium of refractive index $n(x, y, z)$:

$$\int_A^B n(x, y, z)\mathrm{d}l. \tag{1.6}$$

Consider a plane wave travelling through free space. What would happen if half of the wave encountered a glass block of thickness d as shown in figure 1.3?

The time taken to travel through the glass slab

$$\text{time} = \frac{d}{v} = \frac{D_{\text{air}}}{c}. \tag{1.7}$$

In other words,

$$D_{\text{air}} = \frac{c}{v}d = nd. \tag{1.8}$$

Fermat's principle

We know that light travels in a straight line in a region of constant refractive index but if one considers light starting at S in one medium and arriving at P in a second medium, what is the path it will take? Or one could ask what is the path light will take when reflecting in a mirror. In either case, which straight line will it travel along, given the infinite number of choices available? In the seventeenth century, in order to explain refraction, the French mathematician Fermat postulated that light travels between two points on a path which takes the least time. The statement of Fermat's principle has since been restated in terms of optical path length and is able to predict light travel in cases other than refraction. The current form of this principle states that the OPL is stationary with respect to variations of the path and can be described by the equation:

$$\delta \int n(x, y, z)\mathrm{d}l = 0. \tag{1.9}$$

The connection between OPL and Fermat's principle is brought out rather clearly through the following example. In figure 1.4, consider light travelling from point A to B along path L with velocity $v(x, y, z)$, as the refractive index of the medium varies as $n(x, y, z)$. If initially, one considers the time taken to travel a small section $\mathrm{d}l$ of this path, it would be given by

$$\mathrm{d}t = \frac{\mathrm{d}l}{v} = \frac{n\mathrm{d}l}{c}, \tag{1.10}$$

assuming that the refractive index and therefore the velocity are constant over this short distance. Integrating over the entire path would result in the total time of travel:

$$T = \frac{1}{c} \int_A^B n(x, y, z)\mathrm{d}l. \tag{1.11}$$

Path L

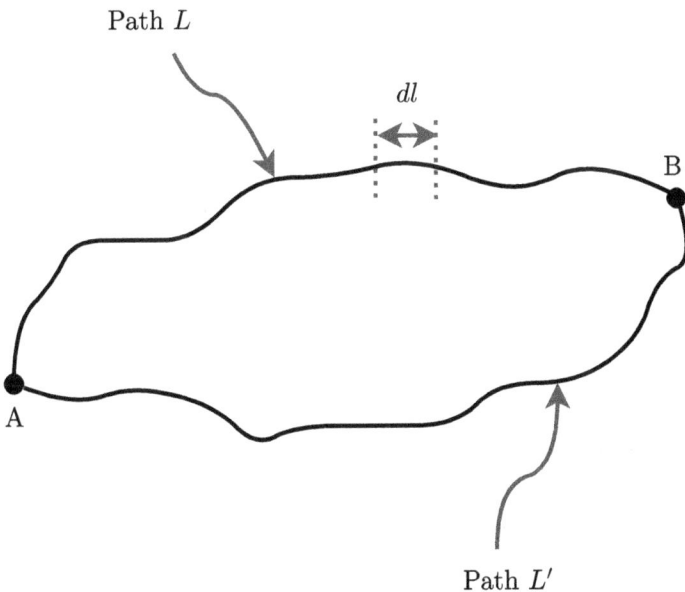

Path L'

Figure 1.4. Demonstrating the connection between OPL and Fermat's principle.

It is clear that the integral part of this equation is equivalent to equation (1.6), which is nothing other than the OPL of the system. Figure 1.4 shows two possible paths (out of infinite possibilities) from A to B. If, as discussed, light travelled on path L, rather than L', it must be clear that the former was the path of least time.

The analogy provided in Feynman's lectures [1] explains this idea very intuitively. Consider a lifeguard standing at point A on the beach. She sees a person drowning at point B in the water and must choose the fastest path to reach them. If the medium was homogeneous and isotropic (a flat beach covered with sand), the fastest path would be a straight line between the lifeguard and the person who needs to be saved. Because of the change in medium, there is an optimum path that will get the lifeguard to the drowning person. This path includes running on sand (the first medium) and swimming in the ocean (the second medium). The path needs to be chosen such that the lifeguard's velocities over the beach and through the water together result in the shortest time overall to cover the distance.

As mentioned earlier, Fermat's principle can be used to predict the path that light will take. Some examples of how this can be done are presented here.

Example 1.1 Using Fermat's principle to prove Snell's law.

How does light travel from point S to P in figure 1.5? We start by writing out the OPL as

$$OPL = n_i SO + n_t OP \tag{1.12}$$

or in terms of time taken,

$$time = \frac{(x^2 + h^2)^{1/2}}{v_i} + \frac{[b^2 + (a - x)^2]^{1/2}}{v_t}. \tag{1.13}$$

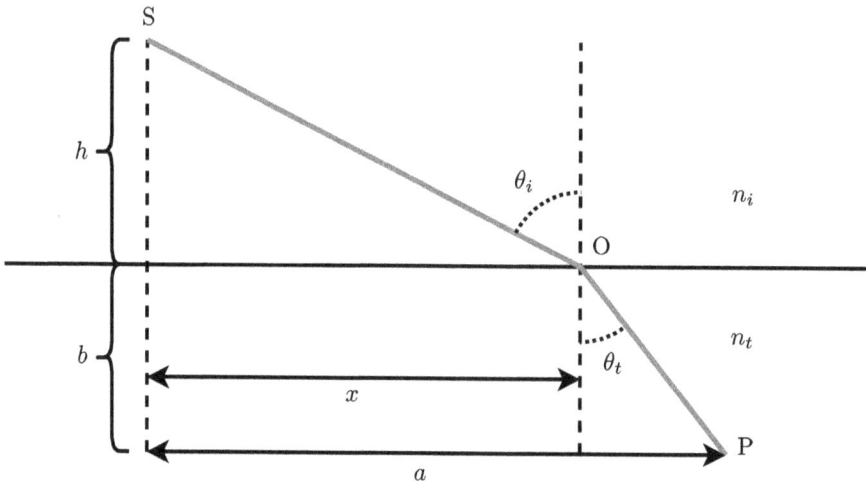

Figure 1.5. Derivation of Snell's law, considering a ray travelling from S in one medium to P in a different medium.

Since we are interested in the path of least time, we need to look at

$$\frac{\mathrm{d}t}{\mathrm{d}x} = 0,$$

which results in

$$\frac{x}{\sqrt{v_i(x^2 + h^2)}} = \frac{a - x}{\sqrt{v_t[b^2 + (a - x)^2]}}. \tag{1.14}$$

This can be written as

$$n_i \sin(\theta_i) = n_t \sin(\theta_t). \tag{1.15}$$

This, of course, is Snell's law, an equation we are all familiar with and which can be rigorously arrived at using wave theory. One can consider Fermat's principle, as a means by which to predict the path light takes without explicitly using its wave nature.

1.2 The wave nature of light

As will be seen in this book, a lot of optical design can be done without using Maxwell's equations [6] or worrying about the wave nature of light. However, there are some phenomena that cannot be explained without employing this facet of light, for example, effects such as interference and diffraction. Neither the existence nor properties of the beams with special behaviour that we will explore in chapter 8 would be easy to explain without the wave nature of light. While geometric optics can be used very effectively to design or predict the behaviour of light under certain circumstances, the wave nature of light provides a deeper understanding of why light behaves the way it does. In fact, rays are related to waves, in that they are the normals to wavefronts.

An alternative way of viewing light is to use the Eikonal approximation, which takes into account some wave properties whilst retaining the simplicity of the ray picture. The wave equation is simplified using the Eikonal approximation by assuming that the amplitude of the wave changes much less rapidly than its phase. The propagation of the resulting wavefronts is then described by the Eikonal equation [7].

1.3 Bridging the gap between theory and design tools

The purpose of this section is to present the ideas discussed in the chapter through the (metaphoric, not optical!) lens of an optical design tool. All relevant chapters of the book will end with such a section. Readers not using these tools can skip this section. However, it is strongly recommended to use a tool, as this really helps in strengthening one's understanding of the fundamental concepts. In this book, we will discuss examples relating to the design software OSLO® (a registered trade mark of Lambda Research Corporation) and Zemax OpticStudio® (a registered trade mark of Ansys), referred to as Zemax henceforth. OSLO stands for Optics Software for Layout and Optimization. Other design tools such as Code V® (a registered trade mark of Synopsis) or LightTools® (a registered trade mark of Synopsis) exist but will not be reviewed. OSLO has an educational version that can be freely downloaded by anyone, after registration. While it is limited in its capabilities compared to the fully paid version, the free version is more than adequate to follow the basic ideas discussed in this book and can even be used to design simple systems. Zemax has also released a freely accessible student version recently.

The following points are true for optical system design tools such as Zemax and OSLO:

- They are spreadsheet-based design tools, with each row (called a surface) representing an optical surface or homogeneous medium.
- Rays are traced through a system using geometric optics. (Note: Newer versions of these tools may include models that allow one to include diffraction effects. In that case, one has to specifically pick the appropriate propagation package. However, these packages will only work under certain conditions.)
- The default travel of light follows what is called *sequential ray tracing*. This implies that a ray can only intersect each surface once and in a specific order (surface #0 followed by #1, #2 and so on). A mirror can be used to change direction.
- Systems can be imaging or afocal ones.
- Different coordinate system options exist. In the local coordinate system, distances are measured from one surface to the next.
- Optimisation and tolerance analysis are possible.

Whichever software is being used, the following steps are required to obtain a final design or solution:

1. System data entry.
2. Ray tracing.
3. Checking optical performance.
4. Optimisation.
5. Tolerancing.

In the various exercises of this book, we will work through steps 1–4 of this process. Tolerancing is a crucial step when designing for industry. It checks the optical quality of the system output, taking into account errors that might arise in the fabrication of individual elements and assembly of the entire system.

Before starting any design, some basic data needs to be entered. They are:

- Aperture stop radius or beam size.
- Field angle.
- Wavelength.

Usually these parameters are part of the system requirements and will not change during the design process, the purpose of which is to find out what elements are required and where they should be located to achieve the goals of the optical system. One starts by entering details of some optical elements and their parameters. Each row of the optical tool spreadsheet therefore may contain the following information:

- Radius of curvature.
- Thickness (distance to next element).
- Aperture radius (size of the element).
- Glass type (material).

However, based on the optimisation, parameters may change to ensure the design meets the desired optical performance. Several tutorials, courses and materials are available online that can be used to learn how to use OSLO [8, 9] or Zemax [10]. The aim of this book is not to introduce the basics of these software tools but rather to employ them to elucidate the concepts being presented.

Now, with light shining brightly, let us forge ahead...

1.4 Problems

1. Using Fermat's theorem, prove the law of reflection.
2. Verify Snell's law using Zemax or OSLO and employing the model shown in figure 1.6. Ensure the incident light is coming from infinity. Vary the value of θ_1 (start at $\theta_1 = 30°$) and obtain the value of θ_2 in each case. The glass type can be chosen as N-Bk7. Compare the values obtained from the software against the values from equation (1.15). Use the value of the refractive index given by the software.

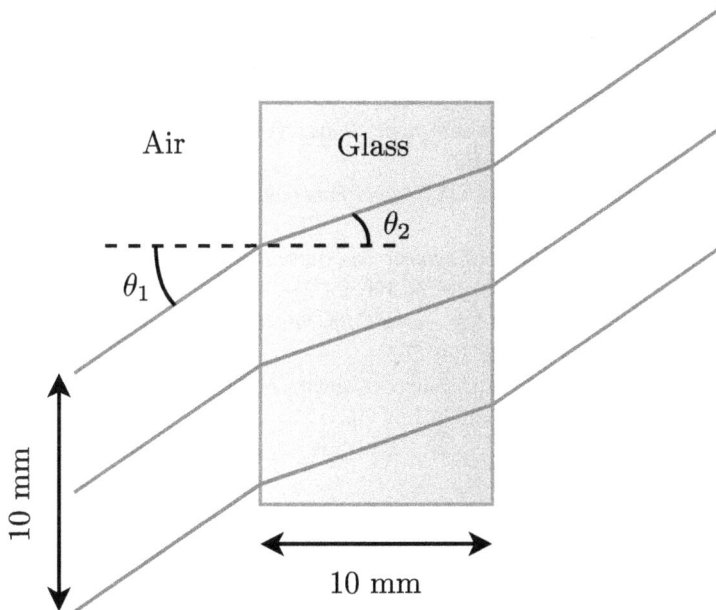

Figure 1.6. Build this model in Zemax or OSLO.

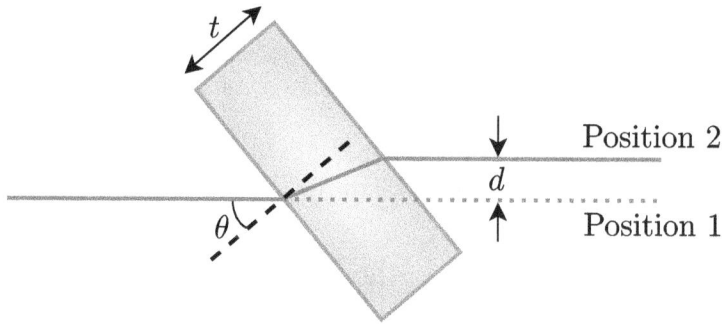

Figure 1.7. Schematic of a refractive index sensor.

3. A beam of a laser light is incident on a beam splitter of thickness t at an angle θ as shown in figure 1.7. The index of refraction of the medium on each side is n and that of the beam splitter is n_1. When the beam splitter is not in place, the light travels to a detector in position 1. When the beam splitter is in place, it travels to position 2. (a) What is the distance d between these two positions? (b) If the beam splitter was replaced with a glass cell, how could this concept be used to measure the refractive index of a liquid?

References

[1] Feynman R P, Leighton R B, Sands M and Hafner E M 1964 *The Feynman Lectures on Physics* **vol I** (Reading, MA: Addison-Wesley)
[2] Hecht E 2012 *Optics* (New York: Pearson)
[3] Fox A M 2001 *Optical Properties of Solids* Oxford Master Series in Condensed Matter Physics (Oxford: Oxford University Press)
[4] Simmons J H and Potter K S 2000 *Optical Materials* (Amsterdam: Elsevier)
[5] Kittel C 2004 *Introduction to Solid State Physics* 8th edn (New York: Wiley)
[6] Shevgaonkar R K 2005 *Electromagnetic Waves* (Electrical and Electronic Engineering Series) (New York: McGraw-Hill)
[7] Rubinstein J and Wolansky G 2004 *Eikonal Functions: Old and New* (Dordrecht: Springer) pp 181–98
[8] Bhattacharya S NPTEL Optical Engineering Course *IIT Madras* https://onlinecourses.nptel. ac.in/noc21_ee81/preview (Accessed: 26 July 2023)
[9] Free OSLO Tutorials *LAMBDA Research Corporation* https://lambdares.com/support-posts/ tag/oslo-examples (Accessed: 26 July 2023)
[10] Free Zemax Tutorials *Ansys* https://support.zemax.com/hc/en-us/categories/1500000770122- Tutorials-Applications (Accessed: 26 July 2023)

IOP Publishing

Introduction to Ray, Wave, and Beam Optics with Applications

Shanti Bhattacharya

Chapter 2

Geometric optics and imaging

2.1 Basic concepts of optical systems

A large part of this book will focus (forgive the pun!) on understanding optical systems overall, as well as how to design and analyse them. For readers more interested in lens design alone, there are a number of excellent books that they can access, starting with the classic by Kingslake [1], as well as several others [2–4]. While optical systems may have a variety of applications, e.g. in sensing [5, 6], light collection, etc, imaging [7] is one of the earliest applications that continues to be required, with more and more demanding specifications. We begin by presenting some ideas that will allow for a simpler discussion of optical systems, particularly those relating to imaging systems. Figure 2.1 can be used to visualise the concepts.

- First, all optical systems can be considered to have a straight line running through them called the optical axis.
- A ray entering the optical system along the direction of this axis will travel through the system undeviated.
- In all circularly symmetric systems, this axis is also the axis of rotation.
- An imaging system has two conjugate points S and P that lie on the optical axis. If the object is located at one, the image will be at the other. They are

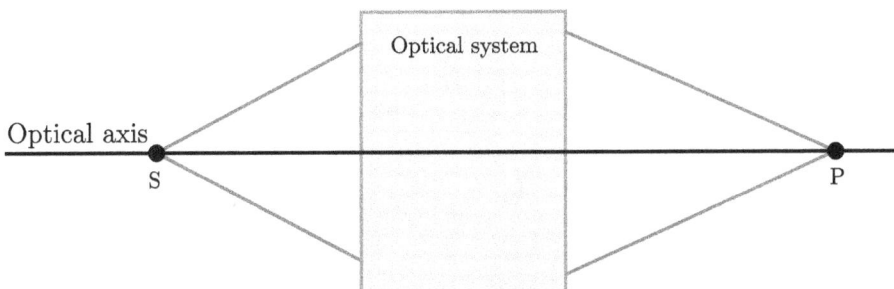

Figure 2.1. S and P are the conjugate points of an optical system.

interchangeable or reversible in optical parlance. It should be noted, however, that while their locations can be exchanged, the quality of the image may vary greatly depending on which point the object is located at.

2.2 What limits imaging?

The primary function of any optical system is to collect as much light as possible from each object point and reshape or focus it. In practical systems, it is impossible to collect all the light. This is demonstrated in figure 2.2. Take the point Q on the tree. Reflected rays, which will be used to form the image, fan out in a cone as indicated by the red arc. However, only the cone of rays (marked in blue) actually make it through the system.

While we avoid the rigorous mathematics of wave theory when dealing with geometric optics, the effects of the wave nature of light still dictate the ultimate behaviour of an optical system. The fact that only part of the wave is captured and the interaction of the captured wave with the finite-sized optics all play a role in the final image quality. We can club these effects under the term *diffraction*. The effects of diffraction are less visible when the wavelength of light λ is much smaller than the physical dimensions of the system. This always breaks down at the edges of an optical system or lens, which is why diffraction can never be completely avoided. Geometric optics works in the region where $\lambda \to 0$. For centuries, it was accepted that the resolution of an imaging system was diffraction-limited. In 1999, Stefan Hell experimentally demonstrated a microscopy imaging technique called stimulated emission depletion (STED), which was able to image better than the diffraction limit. Stefan Hell, Eric Betzig and William Moerner went on to win a Nobel prize [8] for this work in 2014. Newer experimental methods called structured illumination microscopy [9] and computational techniques [10] developed over the last few decades have enabled imaging with an ability better than expected of the conventional optical system.

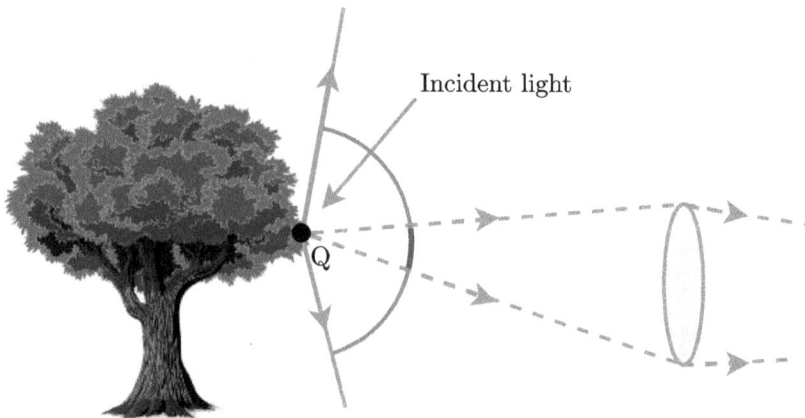

Figure 2.2. Reflected rays form a large cone of light but only a smaller angle (indicated by the blue arc) make it through the system.

Nevertheless, conventional optical systems are still diffraction-limited. Optical microscopes can offer resolutions in the order of hundreds of nanometres. For example, it is possible for confocal microscopes to image features with lateral and axial sizes of 180 nm and 500 nm, respectively [11]. Standard optical microscopes have resolutions in the range from 200 nm to 1 μm, depending on the objective being used. Non-optical systems such as the electron microscope are also diffraction-limited but their de Broglie wavelengths are in the order of picometres and, hence, the resolutions achievable with such systems are far better. Scanning electron microscopes can have resolutions in the range 0.5–4 nm and transmission electron microscopes can have even better values. One disadvantage of such systems is that they need to be operated in a vacuum, which of course means that live cells or samples cannot be imaged. An alternative imaging system with fairly high resolution is the atomic force microscope. It has a lateral resolution of about 30 nm. Although this is not as good as the electron microscopes, the fact that it does not need to operate in a vacuum is an advantage. In one sense, this can be thought of as an optical device, as the signal used for measurement is a beam of light. The light, however, does not interact directly with the sample but reflects a small cantilever that is scanned across and very close to the sample surface. As the cantilever is scanned it deflects, which in turn provides information about the surface topology. The deflection is measured by the amount of reflected light reaching the photo-detector. These different imaging systems have been mentioned briefly here to put the abilities of optical systems into perspective. We now continue with optical imaging systems and the basic theory behind them.

2.3 Refraction at a single surface

Before we can analyse light travelling through an optical system or even a single lens, we need to understand what happens to light when it encounters an interface. We can use our understanding of this interaction as a building block to analyse systems in general. After all, a system can be thought of as a series of interfaces, each with a different shape or curvature and followed by a region of a particular refractive index. In fact, this is exactly how systems are studied with optical design tools such as OSLO^{TM} or Zemax^{TM}. We already know that light at an interface obeys Snell's law. The question of interest at this stage is what shape a surface should take such that light incident on it (from an object) will create an image of that object in the medium after it? One might rephrase this to ask what shape a surface should have to create a focused image. It is pertinent at this point to spend a little time understanding what is meant by focus.

The meaning of focus. In the case of an ideal optical system forming an image, the term *focused image* implies the formation of an image with no aberrations. A collimated beam incident on the same system would result in a diffraction-limited point image at the focal point of the system. Let us use figure 2.3 to help with this discussion. The figure shows two rays from the point object at S travelling to P. We might naively think that if all rays from S arrive at P, we will have a good image. But is this really enough? What property should the rays arriving at P have? Let us follow

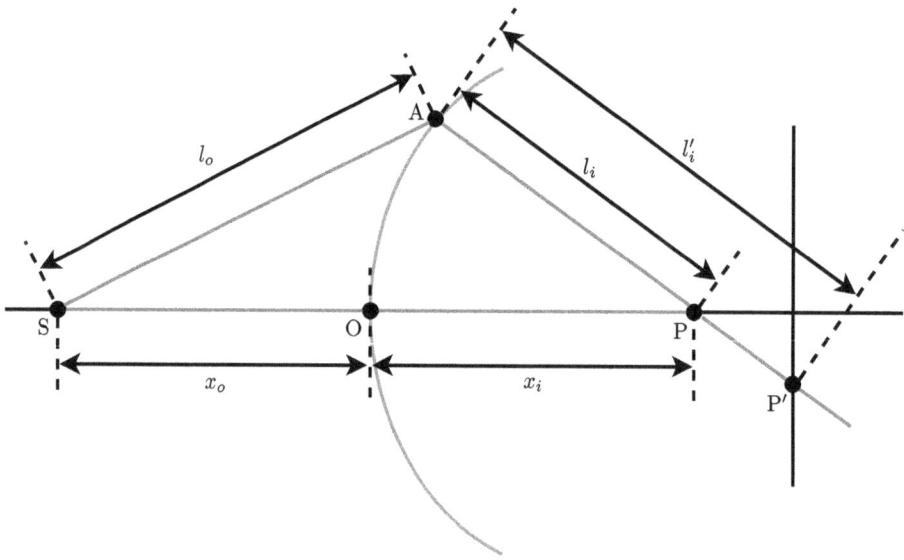

Figure 2.3. Tracing rays across a spherical surface in order to understand what being in focus means.

the rays SO–OP and SA–AP through the system to answer this. If the shape of the surface and the refractive indices on either side are such that

$$n_1 l_o + n_2 l_i = n_1 x_o + n_2 x_i \qquad (2.1)$$

and this is true for all rays travelling from S to P, we can say that surface creates a good image or focuses the light from S at P. Equation (2.1) is the mathematical way of saying the rays arriving at P from S have acquired the same phase or have travelled the same optical path length. This is the true meaning of light being in focus. Imagine that equation (2.1) was not true and instead the equation was given by $n_1 l_o + n_2 l_i' = n_1 x_o + n_2 x_i$. The ray SA–AP' also crosses P but the path length matching condition is not true at this point. The ray needs to travel further to point P' for the condition to be satisfied. So, the mere presence of a ray at a particular location does not imply good imaging or focusing.

Our original question was about the shape a surface needed to have in order to create a good image. We can now answer that. The surface that is needed should satisfy the relation $n_1 l_o + n_2 l_i = n_1 x_o + n_2 x_i =$ constant, thereby ensuring all rays from S are imaged at P. In turns out that for $n_2 > n_1$ such a surface, known as a Cartesian oval, exists. Figure 2.4 demonstrates how the shape of an optical element influences the wavefront incident on it. The central segment of the wave enters the medium n_2 before the edges and, hence, slows down, while the rest of the wave continues to travel with velocity c. This is what causes the wave to bend and in the case of the Cartesian oval *to focus*. By configuring the shape of the element, the wave shape can also be changed, such that it converges, diverges or even has some other desired behaviour.

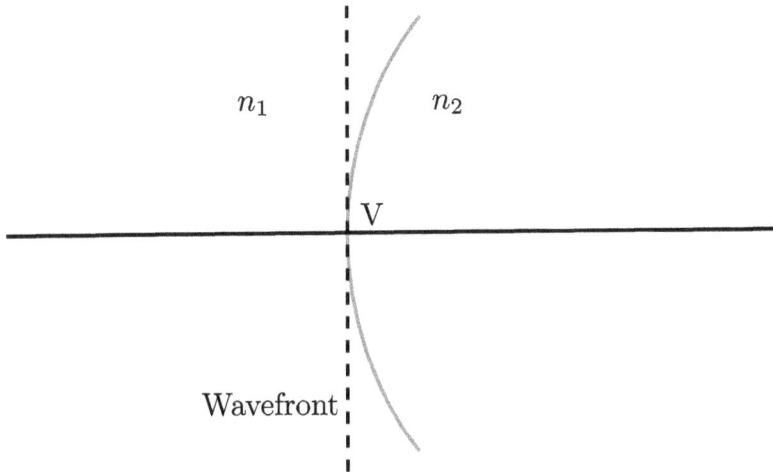

Figure 2.4. A plane wavefront incident on a Cartesian oval. The central part of the wavefront starts to slow down before the edges.

Point to ponder: If Cartesian ovals are the perfect imaging surfaces, why are most optical surfaces spherical-shaped?

We will now try to develop a focusing equation [12], similar to equation (2.1), but for a spherical surface, as that is the shape most optical elements possess.

From figure 2.5, we can write an expression for the OPL of a ray travelling from S to P:

$$\text{OPL} = n_1 l_o + n_2 l_i. \tag{2.2}$$

The distances l_o and l_i are nothing other than the sides SA and AP of triangles SAC and ACP. Therefore, the OPL

$$= n_1[R^2 + (x_o + R)^2 - 2R(x_o + R)\cos\psi]^{1/2}$$
$$+ n_2[R^2 + (x_i - R)^2 + 2R(x_i - R)\cos\psi]^{1/2}. \tag{2.3}$$

The purpose of writing this equation was to arrive at a general expression for any ray travelling through this system. If we pick different rays, all starting from the same point S, then the angle ψ will change, so we use Fermat's principle and derive $d(\text{OPL})/d\psi = 0$. Simplification of this results in

$$\frac{1}{R}\left[\frac{-n_1 x_o}{l_o} + \frac{n_2 x_i}{l_i}\right] = \frac{n_1}{l_o} + \frac{n_2}{l_i}. \tag{2.4}$$

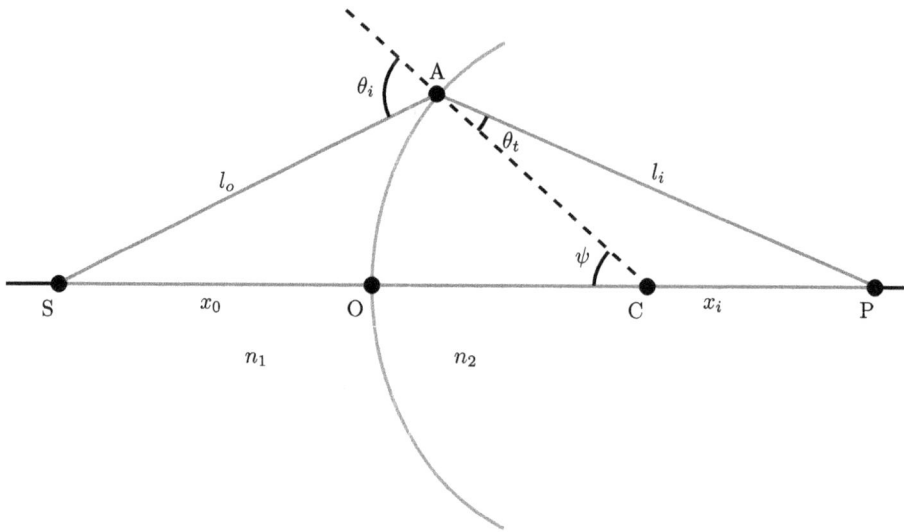

Figure 2.5. Refraction at a single spherical surface. Two rays (in red) are shown travelling from S to P.

In this form, the equation is not very useful because every time we look at a different ray from point S, the point A moves. That is to say, the ray will hit the curved interface at some other location. The only way that equation (2.4) can be satisfied is if x_o and x_i change each time as well, which would make for a lousy imaging system! We need the rhs of equation (2.4) to stay constant for every ray, or in other words, irrespective of the values of l_o and l_i. This can be achieved by approximating the $\cos \psi$ in equation (2.3) to 1, which is valid for small values of ψ. In this case, $l_o \approx x_o$ and $l_i \approx x_i$. Applying this first-order approximation reduces equation (2.4) to

$$\frac{n_2 - n_1}{R} = \frac{n_1}{x_o} + \frac{n_2}{x_i}. \qquad (2.5)$$

Now, the location of the image point depends only on the location of the object point on the optical axis and does not vary depending on which ray is participating in the imaging process. This is true as long as the rays involved are *paraxial* rays, making small angles with the optical axis. The position of the image is solely controlled by the two different refractive indices on either side of the interface and the radius of curvature, which allows one to define a focal length f' of the system as

$$\frac{n_2 - n_1}{R} = \frac{1}{f'}. \qquad (2.6)$$

In itself, this relationship is not useful, as one rarely has an imaging system with different refractive indices on either side of an interface. However, this entire exercise lays the foundation for arriving at the equation for a lens, which can be thought of as two curved surfaces enclosing a medium of a higher refractive index than the air surrounding it.

2.4 Sign convention

The simplest way to define a thin lens is as a refracting system used to shape wavefronts in a controlled manner. At least one surface is curved. Of course, beam shaping, which involves collecting and focusing of light, as well as imaging could also be carried out employing reflection rather than refraction. However, for the moment, we will discuss lenses rather than mirrors. In order to obtain meaningful results, it is necessary to define a sign convention to be used for all the distances involved when arriving at an expression that determines the lens behaviour. There are several existing sign conventions. It is not important which convention one uses, as long as one is consistent. It is assumed that light is travelling from left to right through an optical system. Initially, distances are measured from the vertex formed by the intersection between the first surface and the optical axis:

- Object distances to the left of the vertex are considered positive.
- Image distances to the left of the vertex are negative.
- The radius of curvature (ROC) of a curved surface is taken as negative (positive) if the centre of curvature lies to the left (right) of the vertex.
- Convex lenses have positive focal lengths when light travels from left to right. However, their sign is negative if the direction of travel is reversed.
- Similarly, concave lenses have negative focal lengths when light travels from left to right. Their sign is also changed if the direction of travel is reversed.
- Heights above (below) the optical axis are positive (negative).
- Rays that need to be rotated clockwise to align with the optical axis can be considered to make a positive angle with respect to the axis and vice versa.

2.5 Refraction through a lens

We will apply equation (2.5) to figure 2.6 to arrive at an expression that represents the focal length f of a lens. The basic idea will be to trace a ray through the first surface using this equation and then use the image formed, as the object whose rays need to be traced through the second surface of the lens. Careful use of sign convention 1 is required to obtain an accurate lens equation. In figure 2.6, the shaded intersection region represents a thin lens of thickness d and refractive index n_l surrounded by a medium of index n_s. It is assumed that $n_l > n_s$. The centres of curvature of the two surfaces are located at C_1 and C_2, respectively. An object is situated at S on the axis, at a distance x_{o1} from the first vertex V_1. The purpose of this exercise is to arrive at an expression that will help us locate the conjugate point P in terms of the focal length f of both surfaces. We start by combining equations (2.5) and (2.6). n_1 and n_2 are replaced by n_s and n_m, respectively, and the object and image distances by x_o and x_i by x_{o1} and x_{i1} to obtain

$$\frac{1}{f'} = \frac{n_s}{x_{o1}} + \frac{n_l}{x_{i1}}. \tag{2.7}$$

This expression can be used to locate the image point x_{i1}, given the object position x_{o1}. In figure 2.6, the image formed by the first surface is drawn on the same side as

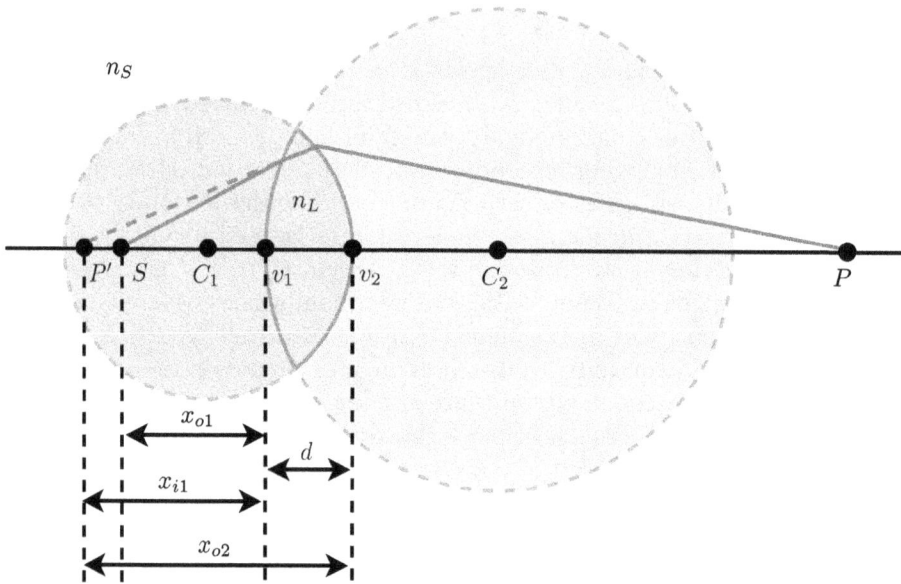

Figure 2.6. Deriving the expression for refraction through a thin lens.

the object. When x_{o1} is large and $\gg f$, x_{i1} is small but positive, which as per the sign convention means the image lies to the right of the surface. As x_{o1} decreases, x_{i1} increases, implying the image moves even further away from the first surface. When the object distance decreases to $x_{o1} = f'$, the image is formed at inf. The object can move still closer to the lens, i.e. $x_{o1} < f'$, but what will happen to the image location? equation (2.7) is satisfied only if $x_{i1} < 0$. This means the image seems to be formed on the same side as the object. In reality the rays would diverge after the first surface and appear to be originating from a point to the left of the vertex. That is why in figure 2.6 the image formed by the first surface is located at P', a distance x_{i1} to the left of V_1.

Point to ponder: It should be clear that if $x_{o1} < f'$, then x_{i1} lies to the left of V_1 but why do we assume this is the object location?

The point to remember is that equation (2.7) gives us the focal length f' of a single curved surface but the main goal is to obtain the focal length f of a single lens. Since the lens has two curved surfaces, we expect it will have a stronger focusing effect than the single surface. In other words, $f < f'$. In that case, it is quite possible that the object distance satisfies the condition $f < x_{o1} < f'$.

With this background, let us consider the refraction that happens at the first curved surface (with ROC R_1) using equation (2.5)

$$\frac{n_1 - n_s}{R_1} = \frac{n_s}{x_{o1}} + \frac{n_1}{x_{i1}}. \tag{2.8}$$

Similarly, the refraction at the second surface is

$$\frac{n_s - n_1}{R_2} = \frac{n_1}{x_{o2}} + \frac{n_s}{x_{i2}}, \tag{2.9}$$

where the object distance x_{o2} is now considered from the vertex of the second surface and is given by $|x_{o2}| = |x_{i1}| + d$.

At this stage, the sign convention is applied. From section 2.4 we know that object distances to the left of the vertex are positive, whereas image distances located on the left are negative. Therefore, $x_{o2} = d - x_{i1}$. Using this and the following assumptions,$R_2 < 0$, $n_s = 1$ and $d \approx 0$, the last one being valid for a thin lens, we arrive at the Lensmaker's formula:

$$\frac{1}{x_{o1}} + \frac{1}{x_{i2}} = (n_l - 1)\left(\frac{1}{R_1} - \frac{1}{R_2}\right).$$

The object and image distances can be replaced by the more familiar symbols u and v, respectively. The constant term on the right relates to the focal power of the system, resulting in the Gaussian lens formula:

$$\frac{1}{u} + \frac{1}{v} = \frac{1}{f}. \tag{2.10}$$

For a lens of refractive index n surrounded by air, the power of the system is defined as

$$P = \frac{1}{f} = (n - 1)\left(\frac{1}{R_1} - \frac{1}{R_2}\right). \tag{2.11}$$

Power is measured in diopters or m^{-1}. It provides a sense of the strength of focusing. A lens with a small focal length, or in other words a tight focus, has a large power. In its more general form, the imaging equation of a lens will be

$$\frac{n_s}{x_{o1}} + \frac{n_p}{x_{i2}} = P, \tag{2.12}$$

if the image is considered to be in a medium of refractive index n_p. In this case, P will be a function of n_s, n_p and n_l.

2.6 Lens imaging conditions for thin lenses

Equation (2.10) can be used to understand the nature of the image given the object position relative to the focal length of the lens. We must keep in mind that the lens equation was arrived at assuming that the lens thickness was negligible, so these results are valid only for a thin lens. The resulting image for each possible object location is specified in table 2.1. It can be seen that as the object moves closer to the

Table 2.1. Object–image relationships for lenses.

Object distance (u)	Image distance (v)	Nature of image with size relative to object size
∞	f	Real point image
$u > 2f$	$f < v < 2f$	Real, inverted, smaller
$2f$	$2f$	Real, inverted, equal size
$2f > u > f$	$> 2f$	Real, inverted, magnified
f	∞	No image formed
$< f$	Same side as object but further from lens	Virtual, erect, magnified

lens, the image moves further away on the other side of the lens. When the object is at the focal length, the image is formed at infinity. Clearly, the image can move no further, were the object to move closer to the lens. For these object distances, the rays from the image will appear to come from a point on the same side as the object. If a screen or camera were placed at this position, no actual image would be captured and hence, the image is called a virtual one.

For thin lenses, a simple ray tracing technique can be used to find the location of an image. There are three rays that can be used to achieve this:

1. A ray travelling through the focal point will emerge parallel to the optical axis after the lens.
2. The ray through the centre of a thin lens passes through undeviated.
3. A ray emerging from the focus will travel parallel to the optical axis after the lens.

Figures 2.7(a) and (b) demonstrate this for the cases when the object distance satisfies $2f > u > f$ and $u < f$, respectively. Triangles SS'O and PP'O from figure 2.7 (a) can be used to calculate the magnification of the system. The magnification is defined as the ratio of the height of the image to that of the object. Taking signs into account, which from the law of similar triangles is equal to

$$M = -\frac{PP'}{SS'} = -\frac{OP}{OS} = -\frac{v}{u}.$$ (2.13)

The magnification gives the size of the image with respect to that of the object. Its value could be less than 1, indicating that the image is smaller than the object.

2.7 Aperture stop, pupils, important rays

Lenses and mirrors have finite sizes. This means that not all the rays reflected off an object make it through an optical system. When the system consists of a single thin lens, the diameter of the lens will determine the number of rays that are used for imaging. Its outer edge can be considered the limiting aperture of the system.

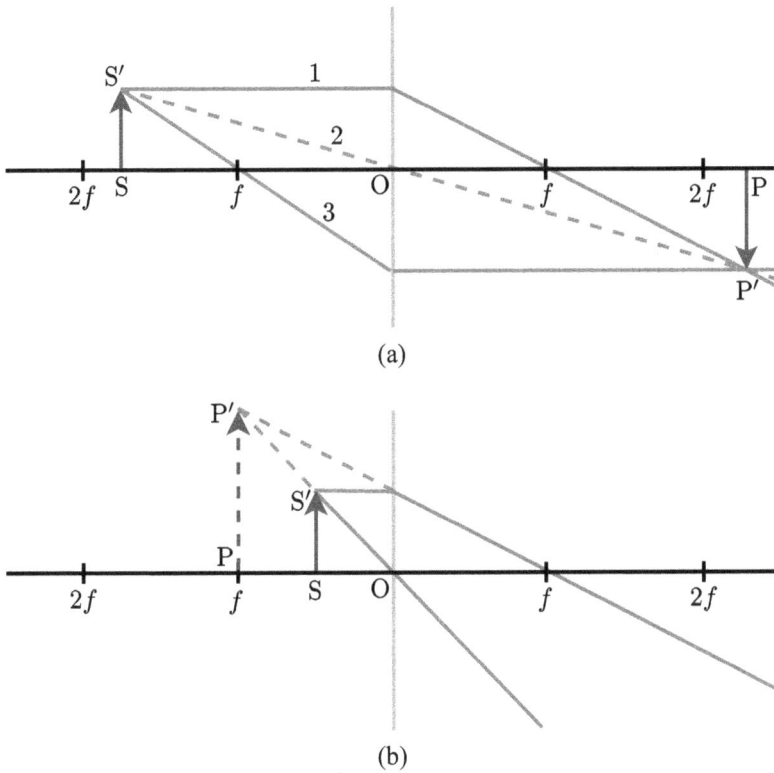

Figure 2.7. Image formation when the object lies (a) between f and $2f$ and (b) at a distance less than f.

However, because elements with power bend light, it is not only the diameter of the element that determines whether a beam makes it through the system or not. A small diameter lens with a large power might steer a light ray with a large incident angle into the optical system. In addition, most optical systems comprise several elements including apertures with no power. One of these elements or apertures will be the bottleneck limiting light through it. A stop is anything that limits which light rays can get through an optical system. There are three common types of stops: aperture stops, field stops and baffles or glare stops. We explore each one of these in turn.

Stop definitions. Consider the optical system shown in figure 2.8, with just two elements: a thin lens and an aperture. At first glance, it may appear that the outer (orange) rays will participate in imaging, as they lie within the diameter of the lens. However, the aperture that is positioned after the lens blocks those rays. Only the smaller cone of (red) rays which make an angle α with each other make it through the system. In this system, the aperture is the aperture stop (AS) of the system. It is important to note that the AS of a system might be a lens or mirror, even if there are simple apertures in the system. The outer rim of every optical element acts as an aperture but only one element acts as the aperture stop.

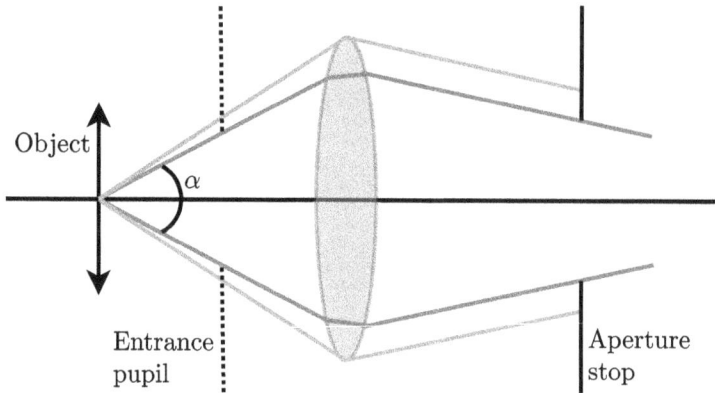

Figure 2.8. Understanding the behaviour of the aperture stop in an optical system.

Optical design tool tip: Rays through the centre of a lens, and through the front and back focal points can always be traced but in reality they may not actually travel through the system.

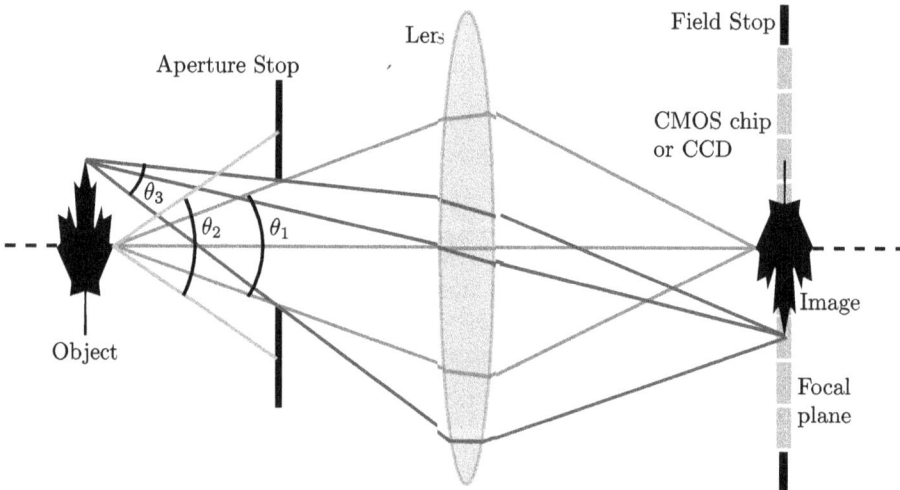

Figure 2.9. Understanding the difference between the aperture and field stops.

Figure 2.9 can be used to further understand the different stops. In this example, an aperture once again acts as the AS of the system but in this case, it lies before the lens. The cone of rays making an angle θ_2 do not make it through the system. As expected, it is the AS that determines the cone of rays (θ_1) that travel through.

Point to ponder: What effect does the size of the cone have on the image?

It should not be hard to imagine that if the rays from the cone θ_2 had travelled through, that image point would be much brighter than the one formed by the rays within the cone θ_1. A larger cone allows more rays to reach the same image point. On the other hand, the field stop determines the extent of the image that is visible. In a digital camera, the active area of the detector would be the field stop of the system.

Apart from affecting the brightness, the AS plays a role in the following important aspects of imaging:

- Depth of focus/field.
- System resolution.
- Certain aberrations.

These points are very important and will be discussed in great depth later. Finally, let us look at the baffle stop. Unwanted light (i.e. light not from the object being imaged) could enter the optical system. Normally, it might not make it through the system but, because of reflection from the sides of the housing as shown in figure 2.10 (a), it is possible that it travels to the imaging plane. Such light would decrease the

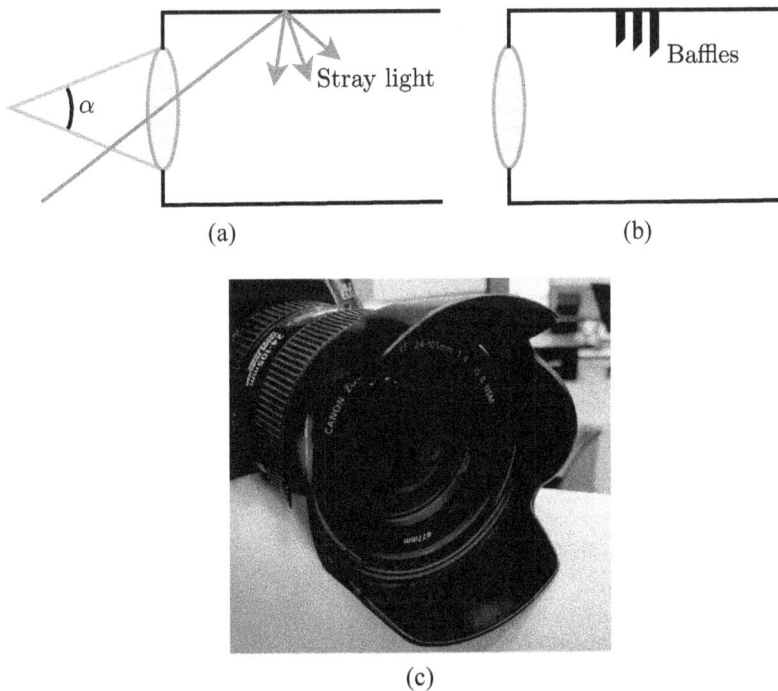

(a)   (b)

(c)

Figure 2.10. (a) Stray light entering a system, (b) the baffles that can prevent it travelling further and (c) a lens hood that acts as a protective shield preventing stray light entering a camera. Photo credit: SB.

contrast and quality of the image. Baffles placed as shown in figure 2.10(b) could break the path of propagation of the stray light and prevent it from travelling further in the system. Alternatively, a lens hood at the front of the system, as seen in the camera in figure 2.10(c) would prevent stray light from even entering the system.

Vignetting. Another point of interest can be gleaned from figure 2.9. If one compares the angles between the rays from an extreme object point and the rays from the axial object point, it should be obvious that $\theta_3 < \theta_1$. Both θ_1 and θ_3 are the cone of rays that make it through the system but that cone itself has a maximum value for the on-axis rays. We have already discussed that the smaller this angle is, the less bright that image point will be. Vignetting refers to the reduced intensity at the edges of an image due to the smaller cone of rays accepted from points further from the axis. Figure 2.11 shows an image of a grey surface with dark corners caused by vignetting of the system. Photographers may sometimes exploit vignetting to draw the viewers' attention to the centre of the photograph.

Pupils. We have seen that an aperture stop is a single element in an optical system. Since this is an imaging system, we can extend the idea of the AS using its image planes. These planes are called the pupils of the optical system and can also be used to determine whether a ray traverses the entire system or not. The *entrance pupil* is the image of the AS as seen by an axial point on the object through the elements preceding the stop. If there are no lenses between the object and the AS, the AS itself will also be the entrance pupil of the system. Similarly, the *exit pupil* is the image of the AS as seen by an axial point on the image plane through the elements after the stop. Because the AS determines the cone of light entering the system, the images of

Figure 2.11. Photograph of a grey surface with the effects of vignetting clearly visible.

the AS also serve the same purpose, determining the rays that enter and leave the system. To better understand this concept, let us work out an example.

Example 2.7.1. An optical system is shown in figure 2.12(a). Find the aperture stop as seen by an object 12 cm in front of the first lens. There are three elements in this system. To find out which is the AS, the cone of light subtended by each one has to be calculated. This is done by assuming each one is the AS, finding the corresponding entrance pupil and calculating the cone of light.

a) We start by finding the image of lens L_2 through L_1, using equation (2.10).

Values	Calculations
$u = -5$	$v = \frac{uf}{u-f}$
	$= \frac{45}{-5+9} = 11.25$ cm
$f = -9$	i.e. virtual image right of lens L

Size of image:

$$M = \frac{-v}{u} = \frac{-11.25}{-5}$$

$$= 2.25 = \frac{y_i}{y_0}.$$

Therefore, the height of the erect image is $y_i = 2.25 \times 2 = 4.5$ cm. We calculate the angle α_1 subtended by this image, as shown in figure 2.12(b),

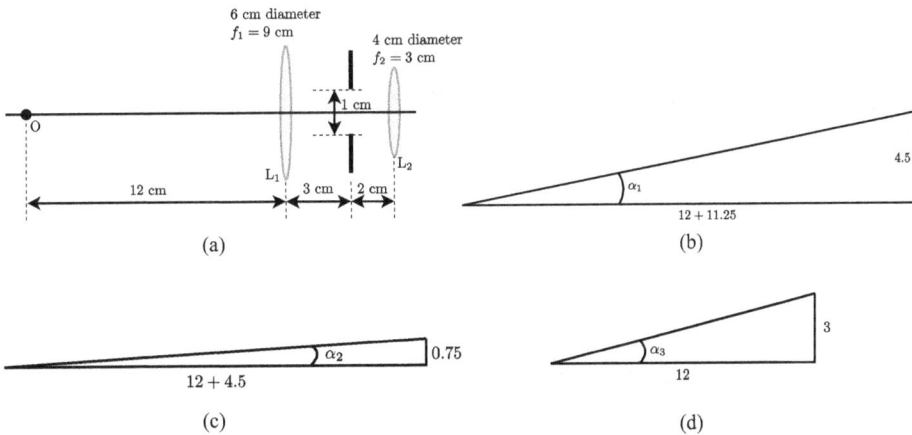

Figure 2.12. (a) Example to demonstrate the method of locating the of AS in a system, (b) showing angle α_1, (c) showing angle α_2 and (d) showing angle α_3.

$$\alpha_1 = \tan^{-1}\left(\frac{4.5}{23.25}\right) = \tan^{-1}(0.195) = 10.95°.$$

(b) Next, we find the image of the diaphragm as seen through lens L_1.

Values	Calculations
$u = -3$	$=\frac{27}{6} = 4.5$ cm
$f = -9$	i.e. virtual image right of lens L_1

Size of image:

$$M = \frac{-v}{u} = \frac{-4.5}{-3}$$

$$= 1.5 = \frac{y_i}{y_0}.$$

Therefore, the height of the erect image is $y_i = 1.5 \times 0.5 = 0.75$ cm and the angle subtended by this image is α_2 as seen in figure 2.12(c):

$$\alpha_2 = \tan^{-1}\left(\frac{0.75}{16.5}\right) = 2.6°.$$

(c) Finally, we calculate the angle α_3 subtended by L_1, as shown in figure 2.12(d):

$$\alpha_3 = \tan^{-1}\left(\frac{3}{12}\right) = 14°.$$

Since α_2 has the smallest value, the diaphragm is the AS of the system.

It is instructive to look at the relative sizes of the different elements in this example. We already have the radius of the entrance pupil (0.75 cm), since we calculated this to arrive at α_2. The radius of the exit pupil can be calculated by finding the image of the diaphragm as seen by lens L_2. It is found to be 1.5 cm. So, what are the final radii that we have?

Radius of initial lens	3 cm
Aperture stop size	0.5 cm
Entrance pupil radius	0.75 cm
Exit pupil radius	1.5 cm

(a)   (b)

(c)   (d)

Figure 2.13. Zoom lens, set at infinite focus (a) entrance pupil, at $f/3.5$, (b) exit pupil, at $f/3.5$, (c) entrance pupil, at $f/22$ and (d) exit pupil, at $f/22$. Photo credit: SB.

We see that the first lens appears relatively large with a radius of 3 cm but of course, the element that determines the amount of light through the system is the diaphragm that has only a 0.5 cm radius. However, this size is not the size of the cone that makes it through, as this is determined by the entrance pupil size which is larger than that of the diaphragm. The fact that optical systems have entrance and exit pupils may not be obvious when looking at the lens of a simple pair of glasses. However, in figure 2.13, a zoom lens that can be attached to a camera is shown. The entrance and exit pupils are clearly visible as bright circles of light. Figures 2.13(a) and (b) show the entrance and exit pupils, respectively, when the lens is set for infinite focus with a 3.5 AS and figures 2.13(c) and (d) are for when the lens is set at the smallest AS setting of 22. In all cases, the lit region is much smaller than the diameter of the lens but what is striking is that even for the same aperture setting, the entrance and exit pupils are of different sizes.

Our ultimate goal is to get to a stage where we can design optical systems with more than one element in them. Part of the process of carrying out such a design is to trace rays through the system and evaluate whether they contribute positively to image formation. While there are an infinite number of rays one can trace, we define some important ones that also aid in the understanding of the entrance and exit pupils.

Ray definitions

The *chief ray* is defined as the ray from an extremity of the object that travels through the centre of the AS. This ray will either travel through or appear to have

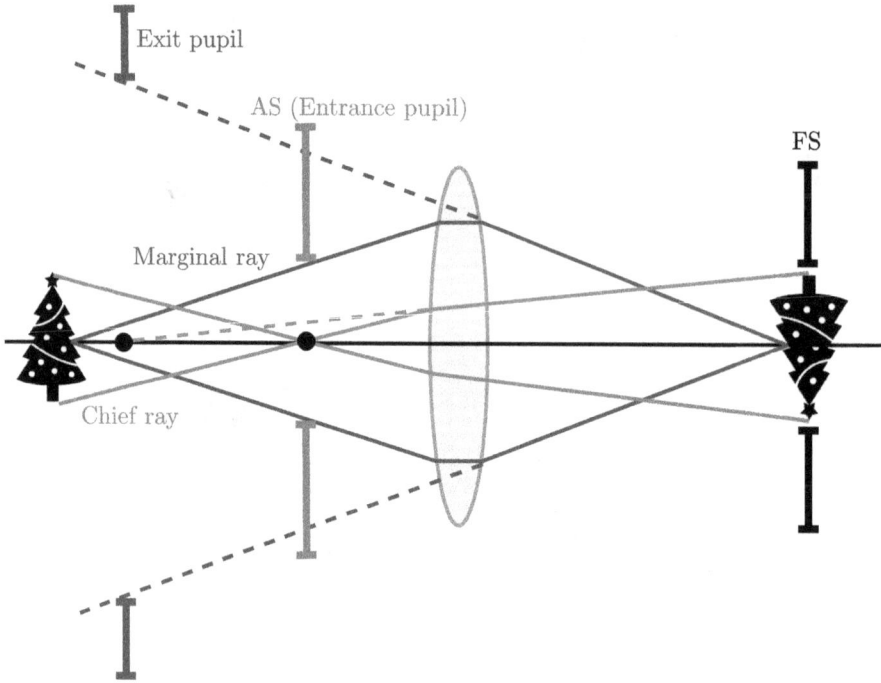

Figure 2.14. A view of the meridional (tangential) plane of an optical system, with marginal and chief rays traced from object to image.

travelled through the centre of the entrance pupil and from the centre of the exit pupil. The *marginal or axial ray* starts from the point that the object crosses the optical axis and travels to the edge of the AS. This ray will either travel to or appear to have travelled from the edge of the entrance and exit pupils. Because of this the height of the marginal ray from the axis at the pupils gives their size. Examples of these rays for a simple system comprising one aperture and one lens are shown in figure 2.14. Both marginal and chief rays are types of meridional rays, which refer to rays that lie in any plane containing the optical axis of the lens. The meridional planes (rays) are also referred to as the tangential plane (rays). The sagittal plane (not shown in this figure) also contains the chief ray but is perpendicular to the meridional plane.

2.8 Important definitions relating to stop size

*f***number**. The parameter called *f*-number or *f/#* compares the focal length of a lens with its diameter. This definition is used for infinite conjugates, i.e. when the object is at infinity

$$f/\# = \frac{f}{D}. \tag{2.14}$$

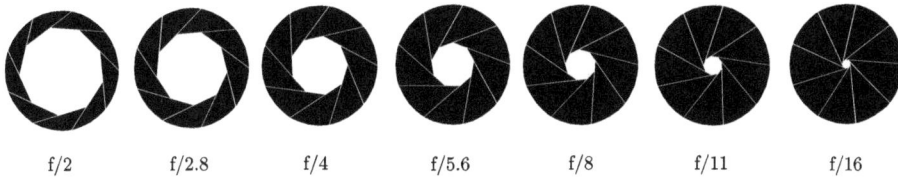

| f/2 | f/2.8 | f/4 | f/5.6 | f/8 | f/11 | f/16 |

Figure 2.15. Different aperture size settings for a system of constant f.

For objects not at infinity, the image distance is used instead of the focal length and this is called the working $f/\#$. The # is actually the value of the f-number itself, so for a lens of $f = 10$ cm and $D = 5$ cm, it would be specified as a $f/2$ lens. The relevance of this term will be more obvious from the other name that is used to refer to it, namely, the speed of the lens. Figure 2.15 helps explain this terminology. It shows a number of apertures for a system of constant f but varying D. The larger the f-number, e.g. $f/16$, the smaller the diameter. Since the clear area of a lens is $\pi D^2/4$, doubling the diameter effectively increases its area (and therefore the light travelling through it) four-fold. A system with a large f-number will need a greater exposure time to capture the image. In other words, a large f-number indicates that the system is slow. Camera stops progress such that 1 stop wider allows twice as much light through, which implies the lens diameter increases by a factor of 1.414. In other words, f-stops vary as a geometric sequence of the powers of $\sqrt{2}$. Cameras can also operate such that the stop is changed by a fractional rather than an integral value, such as a third or a half of a stop, resulting in f numbers such $f/6.3$, $f/6.7$ and so on. The former corresponds to $\sqrt{2}^{5\frac{1}{3}}$ and the latter to $\sqrt{2}^{5\frac{1}{2}}$.

Spot size and resolution. Ideally, the smallest spot size is achieved at the ideal or paraxial image or plane. It will increase the further away one is from the ideal image location. For a collimated beam incident on an imaging lens, the spot size size ϕ, at the focal plane, is related to the diameter D of the aperture stop and is given by

$$\phi = 2.44\frac{\lambda f}{D}. \tag{2.15}$$

It must always be remembered that even a system with no aberration does not form a point image of a point object. Because of diffraction, the best image that can be formed is a circle with a bright central core surrounded by diffraction rings, as shown in figure 2.16. The circle is called an Airy disk and its size is mathematically determined by the Bessel function of the first kind, order 1 [13]. The radius of the first ring of this function is $1.22\lambda f/D$, which accounts for the 2.44 factor in equation (2.15).

Given that a spot is imaged in this way, one can use these diffraction patterns to define the resolution of a system. If there are two points next to each other in object space, the Rayleigh criterion states that they are resolvable when the centre of the diffraction pattern of one point lies directly over the first minimum of the second point, as shown in figure 2.17. The upper rows show the cross section of the pattern for three different cases, whereas the lower row gives the line scan in each case.

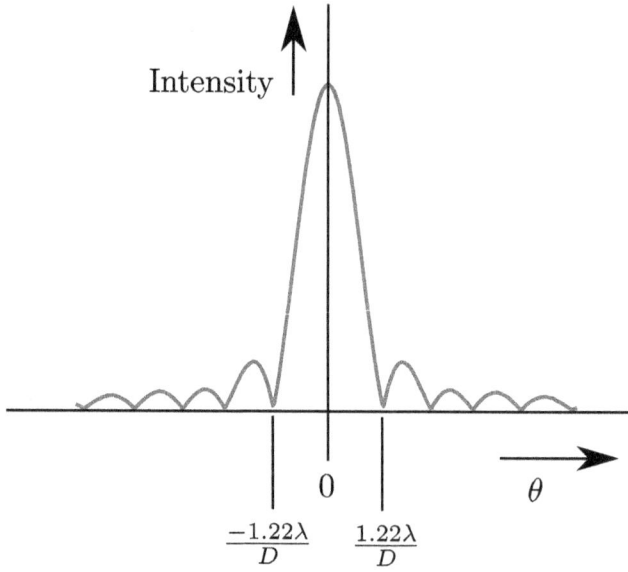

Figure 2.16. Airy rings formed when imaging a point source.

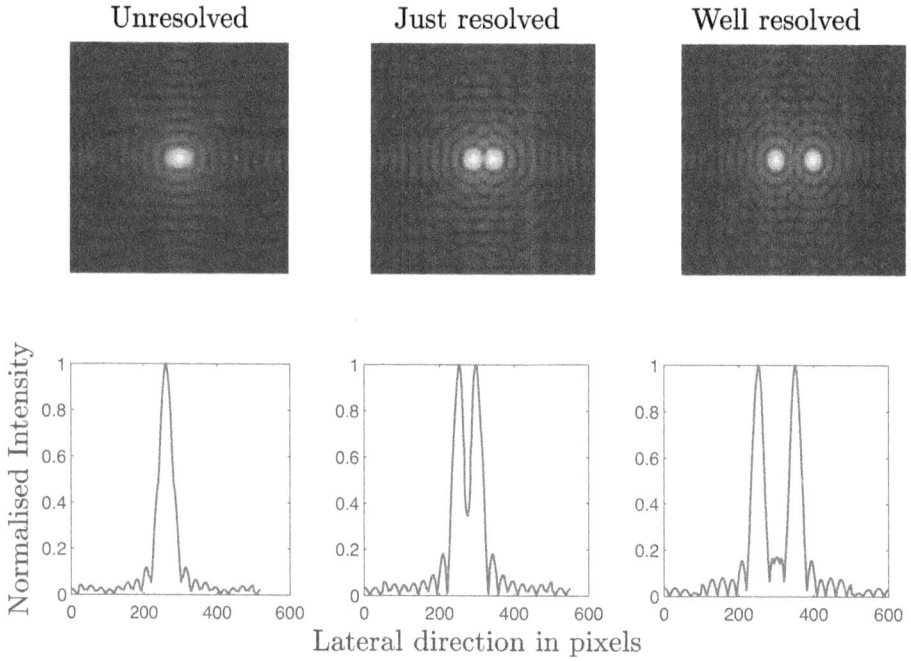

Figure 2.17. Airy rings and their implication on system resolution.

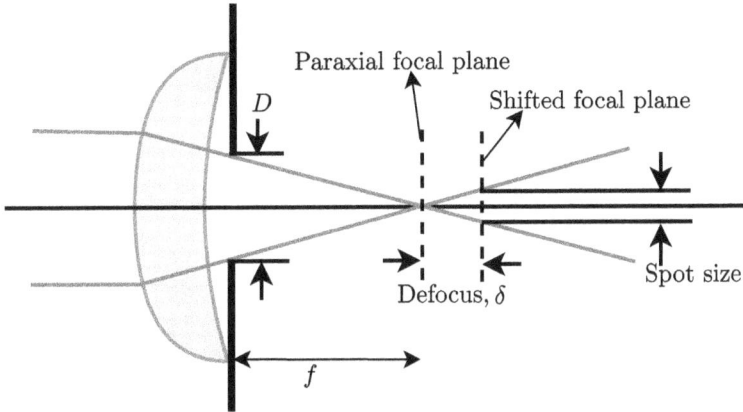

Figure 2.18. The depth of focus of a system is related to the acceptable defocus along the optical axis. Picture credit: MM.

The resolution of a diffraction-limited system is therefore

$$1.22\lambda f/D. \tag{2.16}$$

Depth of focus. If the image plane is not at the paraxial focus, then a point object is not sharply focused but produces a blur spot. The depth of focus (DOF) is defined as the defocus along the optical axis for which the blur is acceptably small. The idea is explained in figure 2.18.

Systems with a larger D have a smaller spot size but also a shorter DOF. These ideas are nicely demonstrated in figure 2.19, where (a) was shot with a f/3.6 setting and (b) with the setting changed to f/8. The latter has a smaller diameter and therefore, a greater depth of focus, which is why the text on the dumbbell in the distance is more legible than in the f/3.6 case.

Numerical aperture. The previous ideas can be brought together with a very important parameter called the numerical aperture (NA) of the system. It is defined as

$$\text{NA} = n \sin \alpha, \tag{2.17}$$

where n is the refractive index of the medium in which the object is and 2α is the angle between the marginal rays that make it through the extremities of the aperture stop, as seen in figure 2.20.

Clearly, a system with a larger NA will form a brighter image. However, the NA plays a role in many important system parameters. To understand all the implications of the numerical aperture, let us assume that the system in figure 2.20 is a single collimating lens and that the object is at the focal position. In that case, if the object were in air, $\text{NA} = D/2f$, where D is the diameter of the lens (and also the entrance pupil, in this case). The spot size [14] defined by equation (2.15) can be rewritten in terms of the NA as

$$\phi = 1.22\frac{\lambda}{2\text{NA}}. \tag{2.18}$$

2-21

(a)

(b)

Figure 2.19. Photos shot with the same camera but at two different aperture settings (a) *f*/3.6 with an exposure time of 1/60 s and (b) *f*/8 with an exposure time of 1/13 s. Clearly, the *f*/3.6 required a shorter time, since it had a larger aperture. Photo credit: SB.

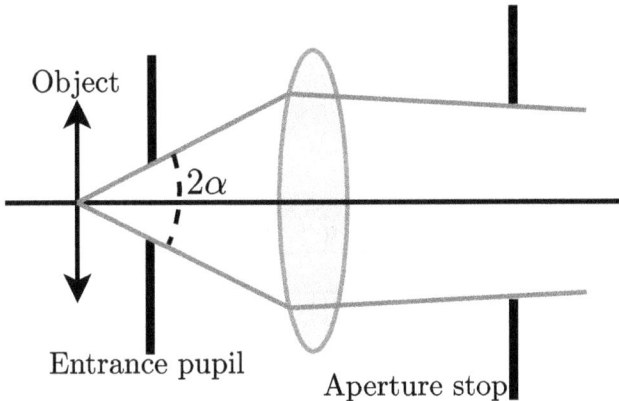

Figure 2.20. Definition of the NA of a system. Picture credit: MM.

At first glance, it would appear that the larger the entrance pupil, the smaller the spot size and to an extent this is true. It must be kept in mind, however, that a larger lens will suffer from more spherical aberration which will degrade the quality of the spot. Equation (2.18) can be thought to represent the geometric or ideal spot size. This is the best size achievable if the system had no aberrations. It is also important to note that the numerical aperture of a system needs to be determined based on its planned usage. Is the system meant for focusing on a collimated beam of light or for imaging a distant object? The numerical aperture will depend on the location of the object.

Point to ponder: Does the spot size depend on the lens or beam diameter?

Field of view. Another important concept is that of field of view (FoV). As the name suggests, it is a term that helps quantify the lateral extent of the object that can be imaged by an optical imaging system. Clearly, this will depend not only on the optics but also on the size of the detector. FoV is defined either as the maximum horizontal, vertical or diagonal size (in units of length) or angular size (in degrees) of the object that can be imaged [14]. Typically, these parameters are measured from the axis. For finite conjugate systems, the full FoV is given by $2 \times h_{\text{obj}}$, as seen in figure 2.21(a). Here, h_{obj} represents horizontal, vertical or diagonal size of the object. For objects at infinity, the full angular field of view (AFoV) is

$$2 \times \text{AFoV} = \tan^{-1}(h_{\text{D}}/f), \tag{2.19}$$

where h_{D} is half the length of the detector, as shown in figure 2.21(b).

Example: Microscope objectives

The objective is a crucial component of a microscope system. It is closest to the object (hence, the name!) and requires fairly sophisticated designing, as it can have

EP

h_{obj}

θ

d

(a)

Thin lens

θ

h_D

f

(b)

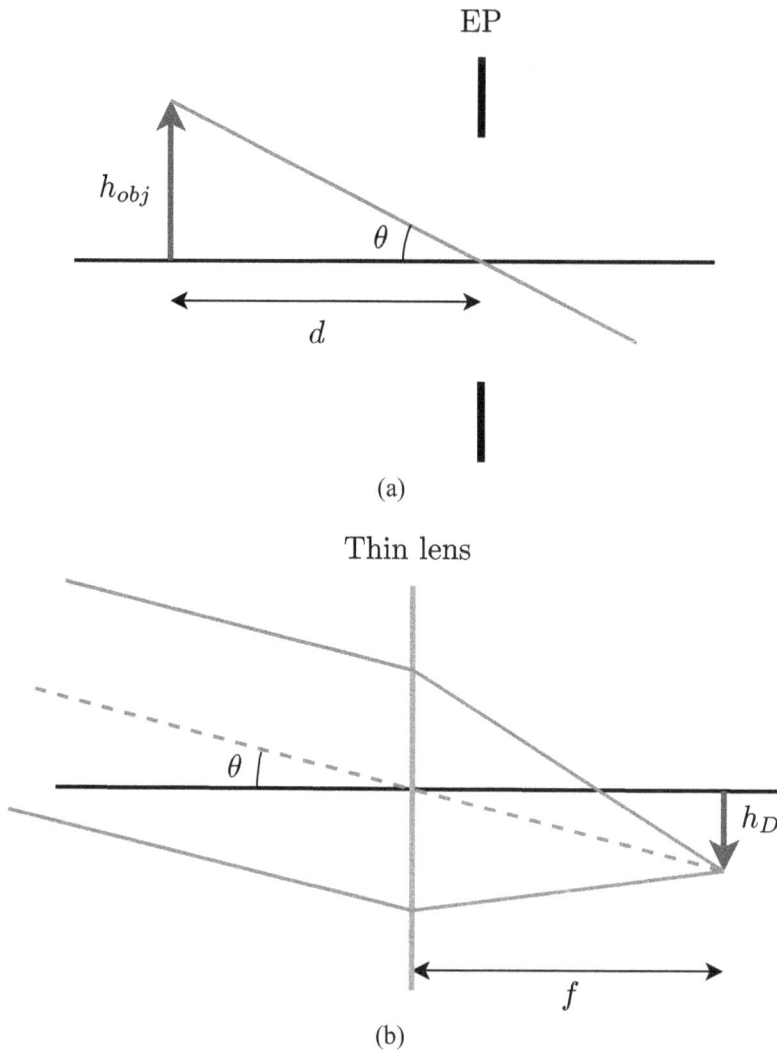

Figure 2.21. Definition of the FoV of a system.

10–15 lenses in it. Objectives are usually specified by their magnification and numerical aperture. Depending on the level of complexity, objectives will correct for most aberrations and also the flatness of the imaging field. There are trade-offs that come into play when choosing the different parameters. Figure 2.22 shows three different objectives with increasing NA and magnification. From these figures, it should be clear that the working distance decreases drastically, as the NA of the objective increases. Theoretically, the highest NA achievable is 1, if the medium between the sample and the objective is air. Higher NAs are achieved using a medium of higher refractive index here. One such class of objectives is called oil-immersion objectives.

Figure 2.23 demonstrates how the change in medium enhances the NA. The sample is prepared on a microscope slide (made of glass) of refractive index n_1. Let

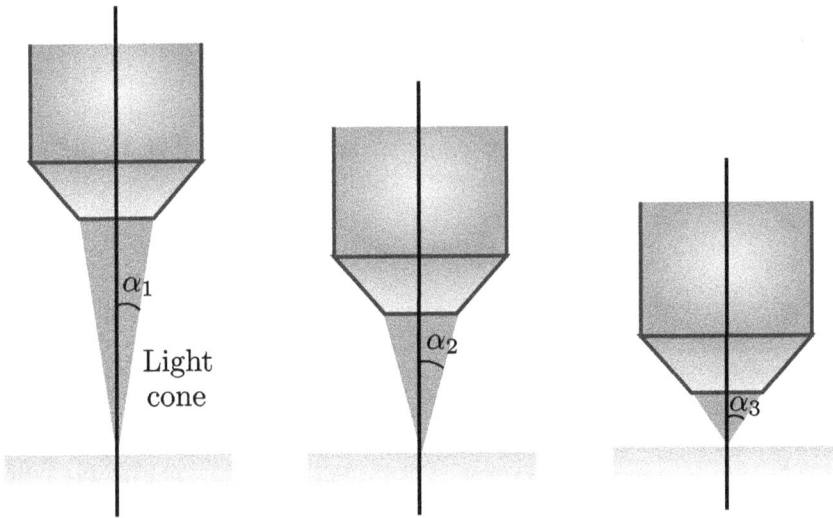

Figure 2.22. Objectives with different NAs. As the numerical aperture increases from left to right, the working distance decreases.

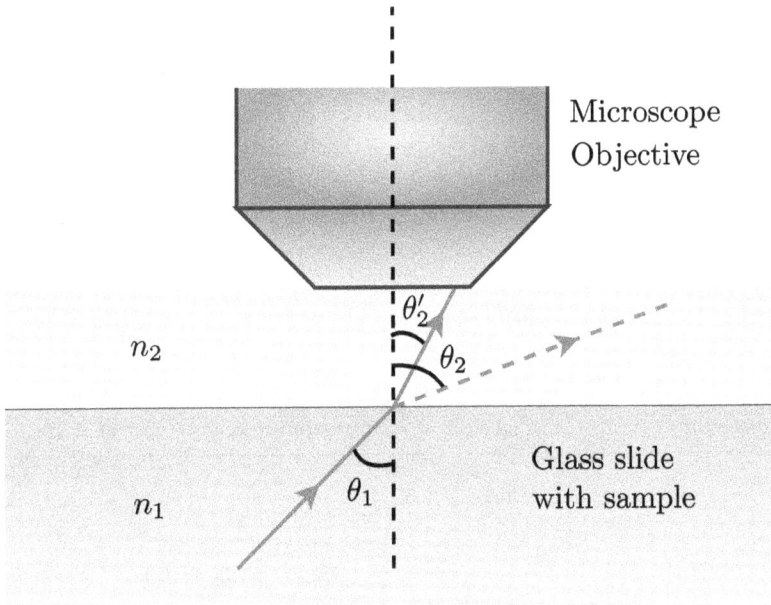

Figure 2.23. Demonstrating the usefulness of an oil-immersion objective.

the medium between the slide and the microscope objective have refractive index n_2. One ray, with an angle of incidence θ_1, is shown travelling from n_1 to n_2. The value of n_2 determines the path that ray takes in the second medium. If $n_1 > n_2$, then the refracted ray makes an angle θ_2 with the normal, and escapes the system. On the other hand, if $n_2 > n_1$, the refracted ray makes angle θ_2' with the normal. This ray travels through the system.

As the magnification of an objective increases, the FoV decreases. The full FoV, in microscopy, can be calculated as (sensor diagonal length)/magnification. For example, if the length of the sensor diagonal is 10 mm, the FoV would be 1 and 0.5 mm for a 10× and 20× objective, respectively.

2.9 Mirrors

Until now, we have considered imaging systems consisting of refractive elements. However, reflective components, such as mirrors, can and are used in several imaging systems. While many of the ideas developed in earlier sections will be valid for mirrors, in this section, we explore the imaging conditions (see table 2.2) and their differences from lenses.

The sign convention defined in section 2.4 remains true except for the following changes:
- Image distances to the left of the vertex are now positive.
- Convex mirrors have negative focal lengths, and concave mirrors have positive focal lengths.

Rays incident on a mirror can be graphically traced using the following concepts:
- Rays parallel to the optical axis reflect such that they seem to originate from the focus of the mirror.
- A ray travelling directly to the focus is reflected parallel to the optical axis.
- A tilted ray heading to the vertex of the mirror is reflected at an equal and opposite angle.
- A ray travelling directly to the centre of curvature is reflected back along itself.

These rules are applied to show image formation by a convex and concave mirror in figures 2.24(a) and (b), respectively.

Table 2.2. Object–image relationships for mirrors.

Object distance (u)	Image distance (v)	Nature of image with size relative to object size
Concave mirrors		
∞	f	Real point image
$u > 2f$	$f < v < 2f$	Real, inverted, smaller
$2f$	$2f$	Real, inverted, equal size
$2f > u > f$	$>2f$	Real, inverted, magnified
f	∞	No image formed
$<f$	$|v| > u$	Virtual, erect, magnified
Convex mirrors		
∞	f	Virtual, point image
$u < \infty$	$|v| < |f|$	Virtual, erect, smaller

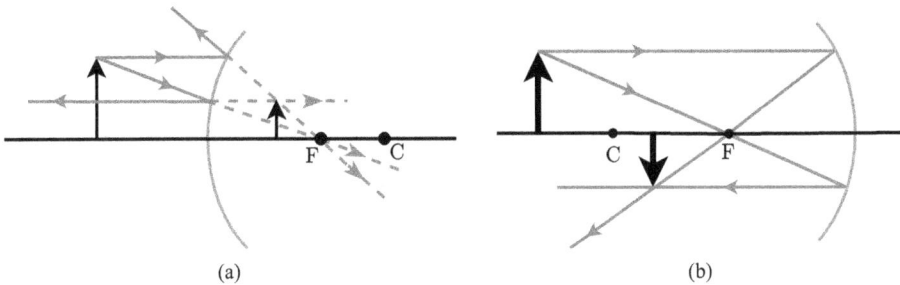

(a)   (b)

Figure 2.24. Image formation with a (a) convex mirror and (b) concave mirror. Picture credit: MM.

Convex mirrors are used in applications that need erect images, such as rear view mirrors of cars. The images are always erect, and smaller in size. They are also useful due to their wide field of view.

Concave mirrors are used to image objects whose distances range from very close by to extremely far away. For example, they are used for imaging teeth by dentists. They are also employed in astronomical telescopes to image distant stars. Apart from being used in imaging, concave mirrors are used in lighting applications such as in car headlamps.

In optics, the terms convex and concave refer to the way light behaves after interaction with such a surface. For example, as we just saw, for *convex* mirrors, a collimated beam of light will appear to focus the incident light after the surface. In the case of *concave* mirrors, the focus will be on the same side as the incident light. While this will always be true for convex and concave mirrors, they can have different shapes, e.g. spherical, elliptical or parabolic.

Parabolic mirrors are very useful for a variety of reasons, which we will explore in chapter 4, section 4.4.4.

2.10 Bridging the gap between theory and design tools

2.10.1 Choosing a surface as the aperture stop of a system

In this section, we look at the implications of choosing a surface as the AS. Optical design tools require the user to set one surface as the AS of the system. There are several ways in which this can be done but the bottom line is that some parameter of, or related to, the AS will be kept constant, irrespective of how big or small other elements are made. This can result in strange ray diagrams, as seen in figures 2.25(a) and (c), if not used carefully!

In figure 2.25, the first surface (on the left) has been set as the AS of the system. This implies that rays will always be traced through this element, as long as they meet other requirements (such as the field angle) of the system. These rays, by very definition of the aperture stop, should reach the image plane. Strange ray diagrams arise if the designer manually enters a diameter of one of the other elements, which is smaller than what it should be. Because this other element should not be limiting the rays, certain software, such as OLSO, will still propagate the rays through non-existent parts of lens and mirrors, which is what is seen in figure 2.25(a). A tool such

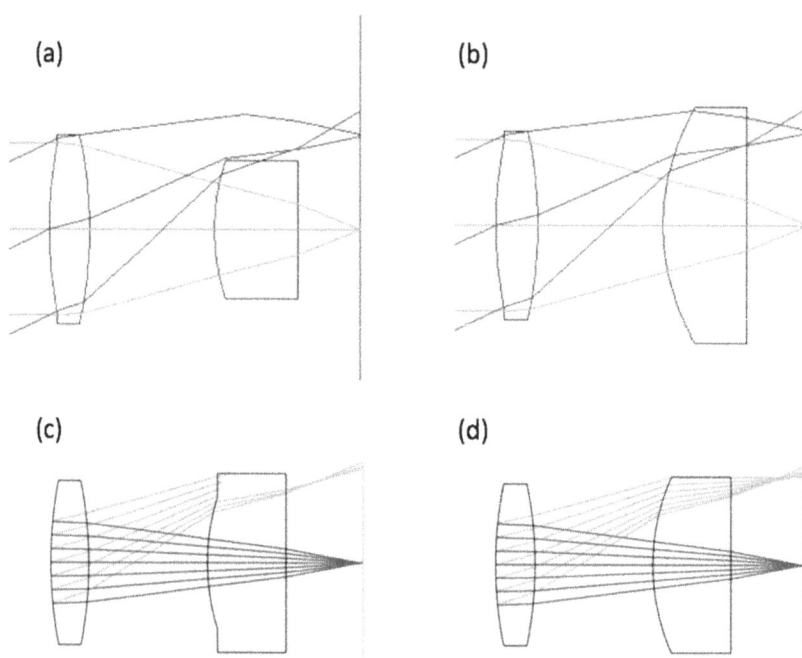

Figure 2.25. Rays traced through an optical system when the first surface is the AS and the controlling parameter is the entrance pupil diameter (EPD). In (a) and (b), OSLO was used and in (c) and (d) Zemax was used. (a) and (c) In these cases, the size of elements after the AS was forced to have a particular size and in (b) and (d) the size was automatically calculated by the software (using in-built solves). Picture credit: SB.

as Zemax, on the other hand, abruptly cuts off those rays, as shown in figure 2.25(c). The correct design procedure is to fix the size of the AS and allow the diameter of all other elements to be set automatically by the software, which is the case for both figures 2.25(b) and (d). Here, we can see that the size of the second lens has been adjusted to make sure the rays (that came through the AS) make it through the second element as well.

All optical design tools require the user to demarcate which element is the aperture stop. Choosing the AS in effect controls the size of the beam travelling through the system. As beam size is an important parameter, design tools offer different ways of assigning a value to it. For example, Zemax provides a variety of options, a few of which are described below:

- *Entrance pupil diameter*

    The aperture stop is defined in terms of the EPD, which is the paraxial image of the aperture stop in object space. The pupil size, rather than the AS size, is to be entered by the user. The beam size will then be governed by EPD, which means that the AS size will be calculated by the software.

- *Float by stop size*

    The size of the AS itself is given by the user. This means that the entrance beam size will change if this parameter is altered. This type is chosen when the stop surface is real and cannot be varied. Choosing this option means that parameters, such as the entrance pupil position, object space numerical aperture and image space $F/\#$, will be calculated automatically.

- *Image space $F/\#$*

    This is the ratio of the paraxial effective focal length calculated at infinite conjugates to the paraxial entrance pupil diameter. If the AS is fixed by this option, then the beam size is controlled by the value of EPD that is calculated from the image space $F/\#$.

2.10.2 Arriving at the correct value for field of view

When designing an infinite conjugate system, one might want to understand what the FoV is given the image sensor details. For example, let us assume that you are using a complementary metal oxide semiconductor (CMOS) camera with 1224×1024 square pixels of side 3.45 μm. The effective focal length of the system is given as 20 mm. What is the field of view of this system? In this case, let us take the diagonal of the camera sensor to be h_{D}, which can be calculated to be 5.506 mm. Substituting this and the focal length into equation (2.19) yields AFoV = 7.84°. This can be entered as the maximum Y field in the field data editor of Zemax or as the field angle in OSLO.

2.10.3 Optimisation

In order to optimise an optical design, one has to enter some parameters, choose the operands of interest and then select the parameters that should or can be varied to enable the system to meet the design requirements. For example, if one was optimising the parameters of a single lens to force it to have a particular focal length, the operand might be the effective focal length (EFL), whereas the variables could be one or both radii of curvature of the lens surfaces. These ideas are explored more in appendix A and problem 9 of this chapter.

2.11 Problems

1. Calculate the radii of curvatures of a biconvex lens made of flint glass of refractive index 1.65 and power of $2D$.
2. What is the focal length of a lens when the medium surrounding it changes from air to a medium with refractive index 1.3? Assume that the material of the lens has refractive index 1.5 and that the original focal length was 5 cm.
3. If instead of changing the medium surrounding the lens, the material of the lens was changed to one with refractive index 1.7, what would the new focal length be? Assume the radii of curvatures are the same as the previous problem.
4. A common occurrence as people age, is that it gets harder for them to focus on nearby objects. This condition is called presbyopia. Assume that an elderly gentleman suffering from this can image objects 150–300 cm away. Because of the accommodation of his eye, what is the range of focal lengths, his eye lens varies over? In such problems, the eye is often represented as a simple lens 1.5 cm from the retina. What would be the focal length of the

spectacles he should wear to create a clearly focused image of an object 25 cm away from his eye? Assume this lens is in contact with his eye lens, when solving this problem. What does this assumption allow us to do when arriving at the solution?

5. When we view a distant object through a lens held at some distance from the eye, the rim of the lens acts like a field stop and the eye itself (specifically, the pupil) is the aperture 7 stop. The lens determines the location and size of the entrance pupil, which in turn determines the AFoV. Imagine that a distant object is being viewed with a 2 cm diameter lens of focal length -17 cm, at a distance of 8 cm away from the eye. Calculate the location of the entrance pupil and use that to find out the AFoV. Show the use of the lens increases the AFoV. To do this, assume a field stop with the same diameter as the lens but with no power.

6. A dentist uses a concave mirror of focal length 3 cm, 1 cm away from a tooth, to see a magnified image of both the tooth and cavity! The cavity is real but is the image real, inverted? With what magnification does the image appear?

7. In an optical system consisting of two identical lenses of focal length 30 cm, separated by 80 cm, find out which element is the AS for an object 50 cm in front of the first lens. Assume the diameters of the lenses to be 20 cm. Draw the elements and entrance and exit pupils. Trace the chief and axial rays through the system.

8. An object is placed in front of a mirror and an image (magnified three times) is formed on a screen. If the object is 5 m from the screen, what can you say about the type of mirror and image formed? What is the focal length of the mirror?

9. *Design goal*: Create a plano-convex lens of focal length 100 mm.

    To start with, ensure light is coming from infinity when carrying out this design. Let the entrance pupil diameter be 10 mm and field angle be 30°. In this example, let the front surface of the lens act as the aperture stop. Assume the thickness of the lens (material N-BK7) to be 10 mm. To start the design process, enter details for any arbitrary plano-convex lens. For example, set the radius of curvature of the first surface of the lens to $R_1 = 50$. It can be seen that the effective focal length of the lens at this stage is 96.4279 mm. This value may be slightly different depending on the lens material and software used. The main point is that the focal length is not 100 mm. Details of how to carry out an optimisation in either Zemax or OSLO (to force the lens to have a particular focal length) are available in table A.1 in the appendix. Note down the radius of curvature of the lens that makes $f = 100$ mm. Use an axial ray height solve to force the image plane to lie at the plane where the ray height is 0. Why is this distance not equal to the focal length?

10. Set up the following system of two N-BK7 lenses. The first lens is a concave lens of radii of curvature -20 and 20 mm, respectively, with a thickness of 5 mm. The second lens is 12 mm away with a thickness of 8 mm. This lens is

(a)

(b)

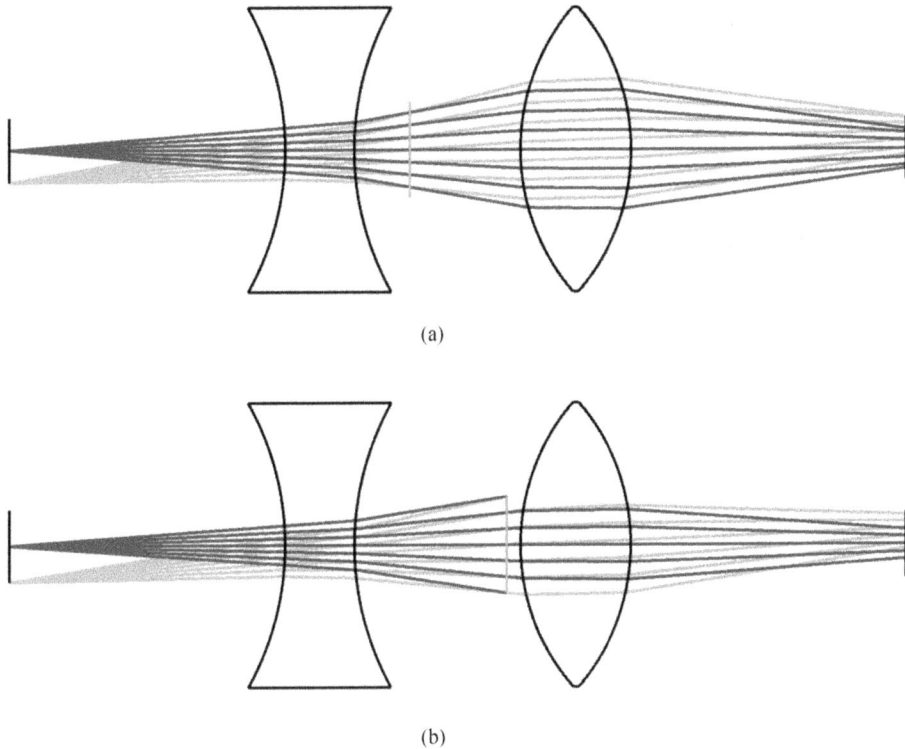

Figure 2.26. Effect of the location of an AS in a system comprising a concave and convex lens separated by a distance of 12 mm. The AS is (a) 4 mm and (b) 11 mm away from the first lens, which impacts the number of rays travelling to the image plane. Picture credit: SB.

a convex lens with radii of curvature 15 and -15 mm, respectively. Make both lenses have a diameter of 20 mm. Set the entrance pupil diameter to be 4 mm and the field angle to be 5°. Place an aperture stop between the lens and use a slider wheel to change its location, keeping the distance between the lenses constant. Figures 2.26(a) and (b) show the layout, when the AS is kept 4 and 11 mm away from the first lens, respectively. What is the radius of the AS in each case? Why does it change? Should the AS not have a constant value?

References

[1] Kingslake R 1978 *Lens Design Fundamentals* (Amsterdam: Elsevier)
[2] Velzel C and Course A 2014 *A Course in Lens Design* Springer Series in Optical Sciences (Dordrecht: Springer)
[3] Laikin M 2001 *Lens Design* Optical Science and Engineering 3rd edn (London: Taylor and Francis)
[4] Geary J M (ed) 2002 *Introduction to Lens Design: With Practical ZEMAX Examples* (Richmond, VA: Willmann-Bell)
[5] López-Higuera J M 2021 Sensing using light: a key area of sensors *Sensors* **21** 6562

[6] Vorathin E, Hafizi Z M, Ismail N and Loman M 2020 Review of high sensitivity fibre-optic pressure sensors for low pressure sensing *Opt. Laser Technol.* **121** 105841

[7] Murphy D B and Davidson M W 2012 *Fundamentals of Light Microscopy and Electronic Imaging* (New York: Wiley)

[8] Hell S W 2014 Nobel Lecture http://www.nobelprize.org/prizes/chemistry/2014/hell/lecture/ (Accessed: 17 June 2023)

[9] Ma Y, Wen K, Liu M, Zheng J, Chu K, Smith Z J, Liu L and Gao P 2021 Recent advances in structured illumination microscopy *J. Phys.: Photon.* **3** 024009

[10] Zheng G, Horstmeyer R and Yang C 2013 A practical algorithm for the determination of phase from image and diffraction plane pictures *Nat. Photon.* **7** 739–45

[11] Fouquet C, Gilles J, Heck N, Santos M D, Schwartzmann R, Cannaya V Y, Morel M P, Davidson R, Trembleau A and Bolte S 2015 Improving axial resolution in confocal microscopy with new high refractive index mounting media *PLoS ONE* **10** e0121096

[12] Hecht E 2012 *Optics* (New York: Pearson)

[13] Arfken G B and Weber H J 2005 *Mathematical Methods for Physicists* 6th edn (Amsterdam: Elsevier)

[14] Greivenkamp E J 2004 *Field Guide to Geometrical Optics* (Bellingham, WA: SPIE)

IOP Publishing

Introduction to Ray, Wave, and Beam Optics with Applications

Shanti Bhattacharya

Chapter 3

Thick lenses

In the last chapter, we analysed lenses, assuming that they were of negligible thickness. This is rarely true for any practical lens. In addition, optical systems often consist of several elements. For example, a camera in a smart phone could have up to six lenses in it [1]. Zoom systems have even more lenses in them, as can be seen in chapter 18 of [2]. In this chapter, we explore the means by which to analyse optical systems with thick lenses (i.e. single lenses whose thicknesses cannot be ignored) and systems with multiple focusing elements or compound lenses in them. We will see that these two cases can actually be explained with one theory.

3.1 Paraxial ray tracing or transfer equations

Ultimately, our goal is to trace rays through an optical system. We look at two transfer equations. One that deals with light travelling through a distance t in free space and the other that takes care of refraction at a curved surface. Figure 3.1(a) shows a ray whose height y_{j-1} and angle u are known at plane $j-1$ on the optical axis. The height of the ray after travelling a distance t is

$$y_j = y_{j-1} + tu. \tag{3.1}$$

In figure 3.1(b), a ray is incident on a curved surface, making an angle i with respect to the normal at that point.

For paraxial rays, Snell's law can be written as

$$n_1 i = n_2 i'$$
$$n_1(\theta + \theta_1) = n_2(\theta + \theta_2). \tag{3.2}$$

But $\theta = y/R = yC$ and the equation can be rewritten, in terms of the curvature C of the surface, as

$$n_1 yC + n_1\theta_1 = n_2 yC + n_2\theta_2. \tag{3.3}$$

doi:10.1088/978-0-7503-5497-4ch3 3-1

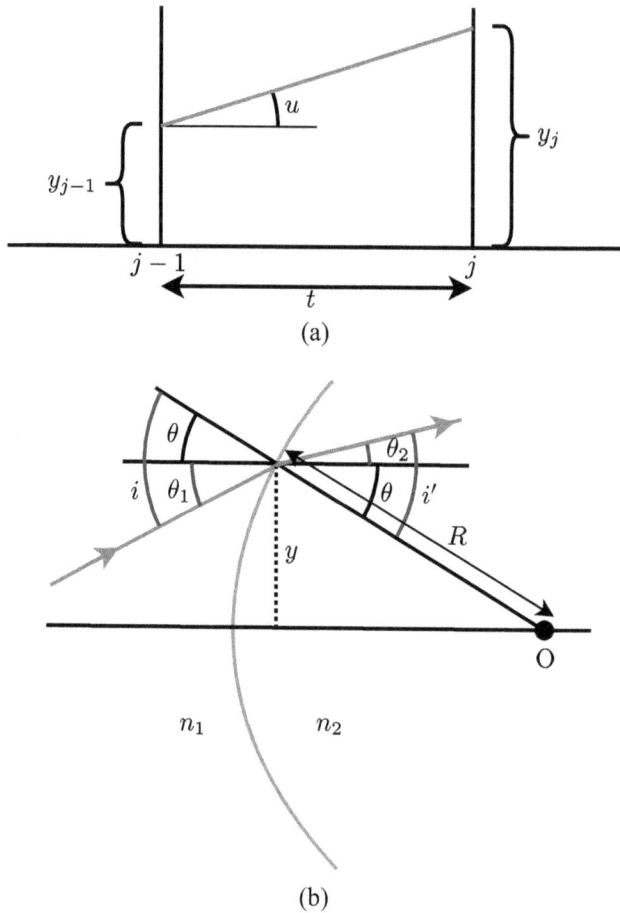

Figure 3.1. Ray tracing (a) through free space and (b) across a curved surface.

The transfer equations can be obtained from equations (3.1) and (3.3). They are given by

$$t = \frac{y_j - y_{j-1}}{u} \tag{3.4}$$

and

$$C = \frac{n_2\theta_2 - n_1\theta_1}{y(n_1 - n_2)}. \tag{3.5}$$

Equations (3.4) and (3.5) are used as *solves* in design tools such as OSLO and Zemax. They are functions whose values will be calculated to meet a specific requirement. Given that the height of the ray is y_{j-1} at a particular surface, equation (3.4) can be used to decide the distance of the next optical element, such that the ray has a height y_j. In equation (3.5), on the other hand, one is solving for the surface curvature, such that the ray has a specific angle after the surface. This is one way of arriving at a lens with a certain focal length.

3.2 Ray transfer matrices

The equations in section 3.1 allow us to calculate the ray height and angle after a surface or at the next surface, given that we have some initial information about the ray at the starting surface. An optical system can be thought of a combination of many different curved and planar surfaces with different refractive indices between them. The ray transfer matrices [3] enable us to trace a ray rather simply through such a system if the following conditions are met:

- The system is circularly symmetric with respect to the optical axis.
- The rays are paraxial.
- The rays are meridional.

We can consider such rays to be completely characterised by their height and angle at a particular location on the optical axis. The act of a tracing a ray means that we can find out its new height and angle at any z position on the axis. This can be understood from figure 3.2, where a ray has height y_1 and angle θ_1 at location z_1 on the axis. After passing through the optical system, the ray parameters are height y_2 and angle θ_2.

We are interested in finding out if there is a way to connect the new ray parameters with the original ones, such as using equations

$$y_2 = Ay_1 + B\theta_1 \tag{3.6}$$

$$\theta_2 = Cy_1 + D\theta_1. \tag{3.7}$$

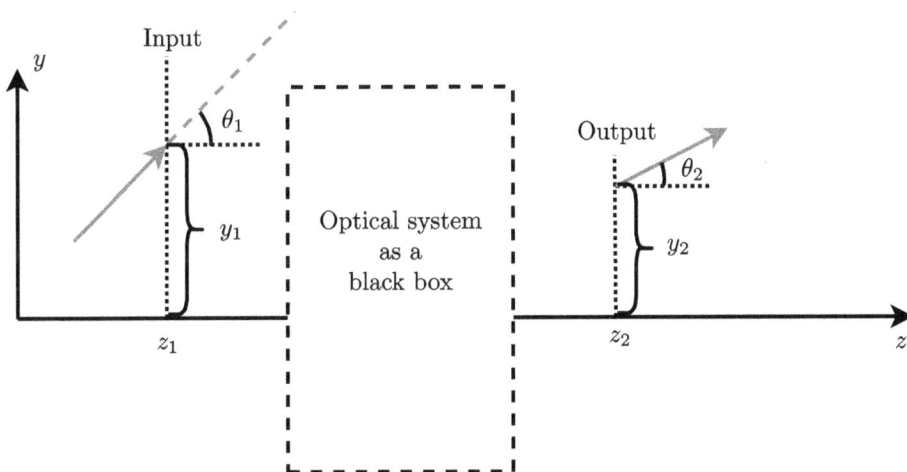

Figure 3.2. A ray is fully defined by its height and angle at a plane.

If such equations existed, they could be written in matrix form:

$$\begin{bmatrix} y_2 \\ \theta_2 \end{bmatrix} = \begin{bmatrix} A & B \\ C & D \end{bmatrix}\begin{bmatrix} y_1 \\ \theta_1 \end{bmatrix},$$ (3.8)

where $\begin{bmatrix} A & B \\ C & D \end{bmatrix}$ is called the *ABCD* or system matrix.

3.2.1 The free space matrix

Using figure 3.1(a) as a base, let us consider the starting and final heights of a ray as y_1 and y_2, respectively, as it travels through a distance t. In free space, the angle of the ray will not change as it propagates. Equations (3.6) and (3.7) will reduce to

$$y_2 = y_1 + t\theta_1$$ (3.9)

$$\theta_2 = \theta_1.$$ (3.10)

Therefore, the *ABCD* matrix for free space is

$$\begin{bmatrix} 1 & t \\ 0 & 1 \end{bmatrix}.$$ (3.11)

The interesting thing to note here is that even travel through free space changes a ray.

3.2.2 Matrix for refraction at a planar surface

Figure 3.3 shows a ray refracting across a planar surface.

Of course, the height of the ray does not change as it crosses the interface but the angle does according to the paraxial Snell's law. This means that $y_2 = y_1$ and $n_2\theta_2 \approx n_1\theta_1$. The resulting *ABCD* matrix is

$$\begin{bmatrix} 1 & 0 \\ 0 & \dfrac{n_1}{n_2} \end{bmatrix}.$$ (3.12)

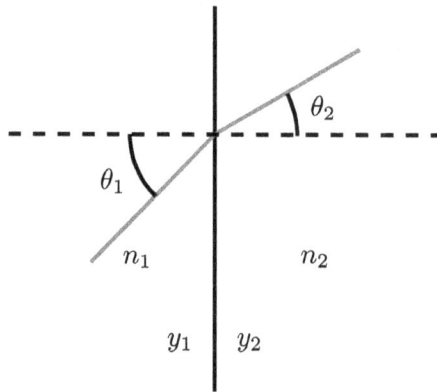

Figure 3.3. Refraction of a ray across a planar surface.

3.2.3 Matrix for refraction at a spherical surface

Figure 3.1(b) can be used to find the system matrix corresponding to refraction at a spherical surface. The ray transfer equations are given in terms of the angles with respect to the optical axis but Snell's law is written in terms of the angles with respect to the normal. Therefore, our interest is in the angle θ_2 in equation (3.2):

$$\theta_2 = \frac{n_1\theta + n_1\theta_1 - n_2\theta}{n_2}.$$

We can replace θ by y_1/R, so that θ_2 becomes

$$= -\frac{n_2 - n_1}{n_2}\frac{y_1}{R} + \frac{n_1}{n_2}\theta_1. \tag{3.13}$$

This results in the $ABCD$ matrix

$$\begin{bmatrix} 1 & 0 \\ \dfrac{-(n_2 - n_1)}{n_2 R} & \dfrac{n_1}{n_2} \end{bmatrix}. \tag{3.14}$$

3.2.4 Matrix for reflection from a planar surface

Once again the height of the ray does not change on reflection from a mirror. The angle has the same magnitude but how do we decide the sign of the angle? We know that angles are considered positive if the rays need to be rotated clockwise (through an acute angle) to reach the optical axis. It might seem like the reflected ray in figure 3.4 will have a negative angle keeping this in mind. However, this convention considers the z-axis to point in the direction of travel of the rays. The reflected rays are travelling from right to left and therefore, this is also the direction of the z-axis for the reflected rays. This means the angle of the reflected ray is also positive. Hence, the $ABCD$ matrix is

$$\begin{bmatrix} 1 & 0 \\ 0 & 1 \end{bmatrix}. \tag{3.15}$$

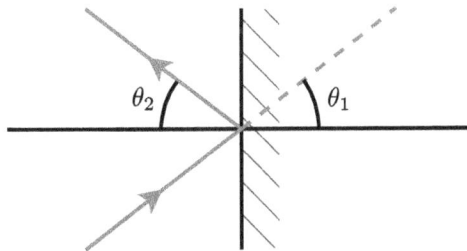

Figure 3.4. Reflection of a ray from a mirror.

3.2.5 Cascaded elements

The power of the matrix method becomes obvious when tracing a ray through more than one optical component. If the various elements have $ABCD$ matrices $M_1, M_2, M_3, \ldots, M_N$, etc, then the system matrix M of the cascaded elements is given by the product of the individual matrices, that is

$$M = M_N \ldots M_3 M_2 M_1.$$

Example 3.1. Find the matrix of a thin lens. The matrix given in equation (3.14) can be used to find the equation or system matrix of a thin lens.

Let us assume the medium surrounding the lens has refractive index n_1 and the lens itself is n_2. The radius of curvature of the surface first encountered by the light is R_1, and the second surface has $\text{ROC} = R_2$. The first and second surfaces will have the matrices M_1 and M_2 associated with them, respectively. Together they represent a lens (thickness = 0), whose system matrix is given by

$$M = \begin{bmatrix} 1 & 0 \\ \dfrac{-(n_1 - n_2)}{n_1 R_2} & \dfrac{n_2}{n_1} \end{bmatrix} \begin{bmatrix} 1 & 0 \\ \dfrac{-(n_2 - n_1)}{n_2 R_1} & \dfrac{n_1}{n_2} \end{bmatrix}. \tag{3.16}$$

The product results in

$$M = \begin{bmatrix} 1 & 0 \\ \dfrac{-(n_2 - n_1)}{n_1}\left(\dfrac{1}{R_1} - \dfrac{1}{R_2} \right) & 1 \end{bmatrix}$$

$$= \begin{bmatrix} 1 & 0 \\ \dfrac{-1}{f} & 1 \end{bmatrix}. \tag{3.17}$$

Equation (3.17) is obtained by recognising the similarity of the C coefficient to equation (2.11). The terms are identical if one replaces the medium surrounding the lens (n_1) by refractive index 1 (i.e. assuming the lens is in air).

This example should demonstrate the power of the matrix method. Every optical operation has an associated matrix, with which one can decipher the effect of that operation on a ray. Given a system matrix, the system can be treated as a black box. As long as the incident ray parameters (height and angle) are known, the parameters of that ray at the output can be calculated without any knowledge of the intermediate optics. However, a system matrix has even more uses, as the next few examples will show.

Example 3.2. Derive an expression for a ray transfer matrix for the system shown in figure 3.5, which is nothing but a thin lens with free space on either side.

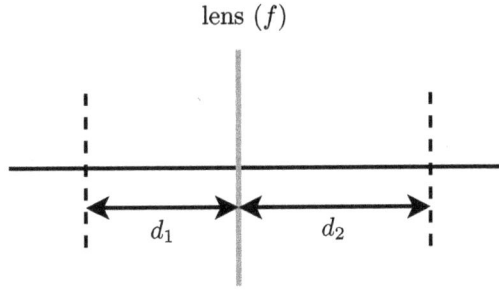

Figure 3.5. Simple optical system.

(a) The system matrix is given by

$$M = \begin{bmatrix} 1 & d_2 \\ 0 & 1 \end{bmatrix} \begin{bmatrix} 1 & 0 \\ \dfrac{-1}{f} & 1 \end{bmatrix} \begin{bmatrix} 1 & d_1 \\ 0 & 1 \end{bmatrix}$$

$$= \begin{bmatrix} 1 - \dfrac{d_2}{f} & d_1 - \dfrac{d_1 d_2}{f} + d_2 \\ \dfrac{-1}{f} & \dfrac{-d_1}{f} + 1 \end{bmatrix}. \qquad (3.18)$$

(b) What does it imply if the system matrix satisfies this condition: $\frac{1}{d_2} + \frac{1}{d_1} = \frac{1}{f}$?
Consider the B coefficient of the resultant of (3.18).
If $\frac{1}{f} = \frac{1}{d_2} + \frac{1}{d_1}$, then $B = 0$ and the system matrix reduces to

$$\begin{bmatrix} 1 - \dfrac{d_2}{f} & 0 \\ \dfrac{-1}{f} & \dfrac{-d_1}{f} + 1 \end{bmatrix}.$$

If the starting height of a ray incident on this system was y_1, its height after the system (i.e. at distance d_2 from the lens) would be

$$y_2 = \left(1 - \frac{d_2}{f} \right) y_1.$$

Point to ponder: This system is shown in figure 3.6. What is the significance of the ray height being independent of θ_1?

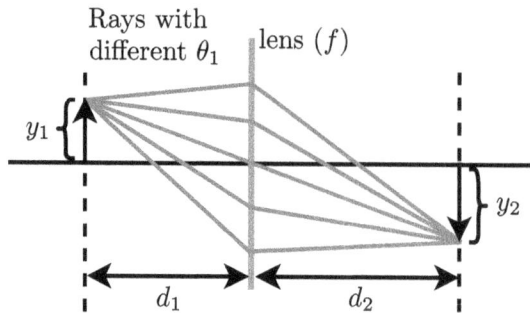

Figure 3.6. Rays with the same object height but different angles.

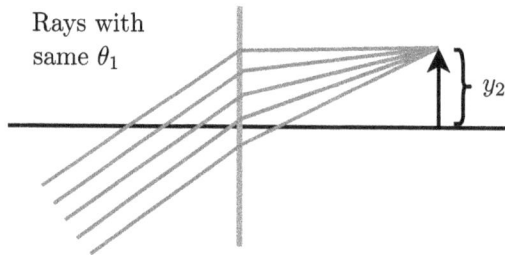

Figure 3.7. Rays with the same angle but different heights.

(c) Let us consider the case when d_2 is chosen such that $d_2 = f$. In this case, the A coefficient of (3.18) reduces to 0. or in other words $y_2 = B\theta_1$. This means all rays with incident angle θ_1 arrive at a point with height y_2 irrespective of their initial height, as shown in figure 3.7.

3.3 The Lagrange invariant

The determinant of any valid optical system matrix is found to always be the ratio of the incident to the emerging refractive index, i.e. n/n'. This can be verified by looking at any of the matrices in section 3.2. Most of them have determinant 1, as the systems were surrounded by air. This fact can be used to arrive at an extremely powerful conservation law called the Helmholtz, the optical or the Lagrange invariant [2, 4, 5].

3.3.1 Arriving at the Lagrange invariant: method 1

To derive this, let us assume that two rays with height and angle coordinates given by y_{a1}, θ_{a1} and y_{c1}, θ_{c1} are allowed to travel through a paraxial optical system with system matrix $ABCD$. At this point, the system matrix is unknown but the output coordinates of the traced rays can be obtained and are given by

$$y_{ak} = Ay_{a1} + B\theta_{a1}$$

$$\theta_{ak} = Cy_{a1} + D\theta_{a1} \qquad (3.19)$$

and

$$y_{ck} = Ay_{c1} + B\theta_{c1}$$

$$\theta_{ck} = Cy_{c1} + D\theta_{c1}. \qquad (3.20)$$

Tracing a pair of rays in this manner allows us to calculate the unknown $ABCD$ values. Using the previous equations, we obtain

$$A = \frac{y_{ck}\theta_{a1} - y_{ak}\theta_{c1}}{y_{c1}\theta_{a1} - y_{a1}\theta_{c1}} \qquad (3.21)$$

$$B = \frac{y_{ak}y_{c1} - y_{a1}y_{ck}}{y_{c1}\theta_{a1} - y_{a1}\theta_{c1}} \qquad (3.22)$$

$$C = \frac{\theta_{ck}\theta_{a1} - \theta_{ak}\theta_{c1}}{y_{c1}\theta_{a1} - y_{a1}\theta_{c1}} \qquad (3.23)$$

$$D = \frac{y_{c1}\theta_{ak} - y_{a1}\theta_{ck}}{y_{c1}\theta_{a1} - y_{a1}\theta_{c1}}. \qquad (3.24)$$

This derivation is only valid as long as the two rays traced are linearly independent of each other. It is common to use the axial and the chief rays for this purpose, which should explain the choice of subscript labels. Beyond the fact that they meet the mathematical requirements of the derivation, they also provide vital information of the system through the Lagrange invariant. Careful perusal of equations (3.21)–(3.24) shows that in all cases the denominator equals

$$LI = y_{c1}\theta_{a1} - y_{a1}\theta_{c1}. \qquad (3.25)$$

Since the determinant $|AB - CD|$ of the system is a constant, this implies that LI is also a constant.

3.3.2 Arriving at the Lagrange invariant: method 2

We can arrive at the same result in a slightly different way. Let us trace the axial and chief rays across a surface using equation (2.12). We use the optical system of figure 3.8 to help visualise this. The system consists of two thin lenses and an aperture.

Consider the image formation at the first thin lens. We can rewrite the object and image distances as $x_{o1} = y_{a1}/\theta_{a1}$ and $x_{i2} = y_{a1}/\theta_{ak}$ assuming the axial ray height at that lens is y_{a1}. Therefore, the equation for the axial ray will be

$$\frac{n_s\theta_{a1}}{y_{a1}} + \frac{n_p\theta_{ak}}{y_{a1}} = P. \qquad (3.26)$$

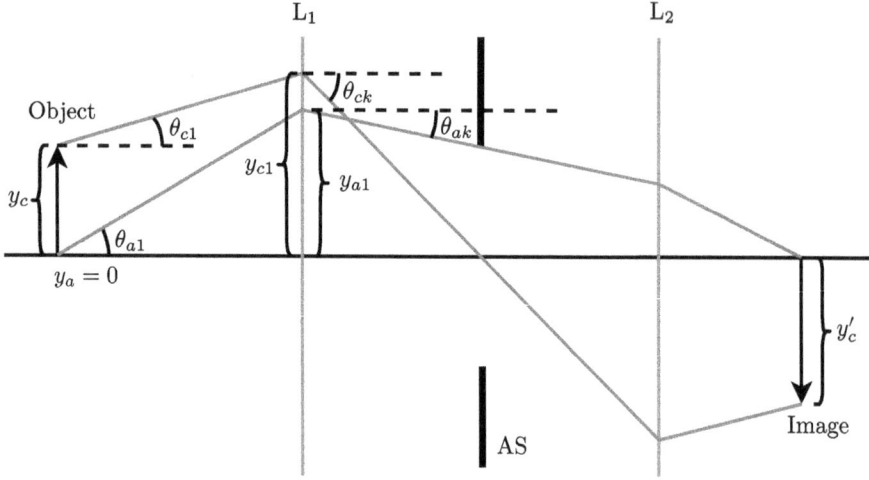

Figure 3.8. Deriving the Lagrange invariant by tracing an axial and chief ray through an optical system. The object and image heights are y_c and y_c', respectively.

Similarly, the equation for the chief ray, given the object height is y_c, will be

$$\frac{n_s \theta_{c1}}{y_{c1} - y_c} + \frac{n_p \theta_{ck}}{y_{c1} + y_c'} = P \tag{3.27}$$

Since the power P of the lens is the same in both cases, we can equate these equations and obtain the following equality:

$$n_s \left(\theta_{a} y_{c1} - \theta_c y_{a1} \right) - n_s \theta_{a1} y_c = n_p \theta_{ck} y_{a1} \frac{y_{c1} - y_c}{y_{c1} + y_c'} - n_p \theta_{ak} (y_{c1} - y_c)$$

Or in other words, since the right hand side goes to zero

$$n_s \left(\theta_a y_{c1} - \theta_c y_{a1} \right) = n_s \theta_a y_c, \tag{3.28}$$

The right hand side of equation (3.28) is the LI at entrance pupil and the left hand side represents the LI at the object plane. We can generalize this and write the invariant term in terms of the incident axial and chief rays as

$$\text{LI} = n \left(y_c \theta_a - \theta_c y_a \right). \tag{3.29}$$

The incident medium has refractive index n in this case. This quantity LI does not change as the ray traverses an optical system. Since LI is conserved, if we find its value at one plane of the system, we can use that to give us information about some other point of the system. Let us look at the object plane, where the axial ray height $y_a = 0$. From equation (3.29), the LI reduces to $n y_c \theta_a$. Going back to the ideas introduced in section 2.8, this can be written as

$$\text{LI} = n \theta_a y_c = \text{NA}h, \tag{3.30}$$

where $n\theta_a$ is the numerical aperture of the paraxial system and $y_c = h$ is the height of the object.

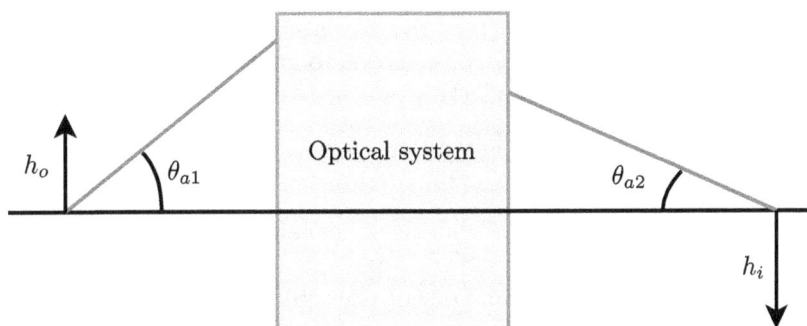

Figure 3.9. Working out problems with the help of the LI.

Example 3.3. Using the LI to obtain the height of the image

Consider an optical system surrounded by air, as shown in figure 3.9.

If the axial ray angles in object and image space are known and given by $\theta_{a1} = 0.033$ radians and $\theta_{a2} = -0.04755$ radians and the object height h_0 is $+20$ mm, by using equation (3.30), we can calculate the image height h_i:

$$h_i = h_o \frac{\theta_{a1}}{\theta_{a2}} = -14.0187 \text{ mm}. \tag{3.31}$$

The image height has been calculated without explicitly needing to find the location of the image.

Example 3.4. Using the LI to evaluate the information-carrying capacity of the system.

In order to employ the idea of spot size as shown in equation (2.18), we multiply and divide equation (3.30) by $2\lambda(1.22)$ which yields

$$\text{LI} = \frac{0.61 \lambda h}{r}, \tag{3.32}$$

where the resolution r of the imaging system is given by equating to the spot size ϕ in equation (2.18).

This brings up a very interesting and important point. In the previous chapter, we talked about how much light makes it through a system and also about the spot size, which determines the resolution but we never explicitly discussed the information-carrying capacity of an optical system. We have just arrived an expression (equation (3.30)) for the Lagrange invariant in terms of the chief ray height and the axial ray angle. The former relates to the field of view of the system, whereas the latter to its numerical aperture or resolution. Another way to think of this is that axial ray angle relates to the bandwidth of the optical system. One can therefore think of the LI as determining the amount of object information that makes it through the system. An important, if subtle point here is that the information is conserved as the light travels through the system. If we imagine that the full field size is $2h$, then given the resolution of either equation (2.16) or equation (2.18), the number of spots N_S needed to image the full field will be $2h/r$. Substituting this into equation (3.32)

$$LI = \frac{0.61 \lambda N_S}{2} \approx \frac{\lambda N_S}{4}. \tag{3.33}$$

So, the Lagrange invariant also helps us understand the maximum information-carrying capacity of an optical system. To put this into perspective, consider that you want to buy a digital camera and one of your criteria is the number of pixels the sensor has. The LI tells us that a larger number of pixels on the sensor do not necessarily imply better resolution, as it is the optics of the system that determines that. Consider that you are buying a 100 MP camera. While sensors are not usually square in size, let us assume that this means 10^4 pixels along one direction. This means the optical system must be able to resolve 10 000 points in one axis in order for you to obtain the full benefit of this sensor!

3.3.3 Principal planes

In earlier sections, we saw how the use of matrices greatly simplified the act of tracing rays through an optical system. However, understanding the imaging and focusing of such a system at first glance does not seem anywhere as easy as it was for a single thin lens. It would be extremely useful if there was a similar way to analyse multi-element systems. This is where the principal planes come into the picture. They are perpendicular to the optical axis and run through the principal points that will be derived in this section. We will understand their significance by answering this question: 'Is there some way to analyse a system as if it were a thin lens?' The answer would be yes if we could represent a system matrix in the form of equation (3.17). If the lens was considered to have power P and had media of refractive index n and n' on either side of it, the general form of the equation would be

$$M_{HH'} = \begin{bmatrix} 1 & 0 \\ \dfrac{-P}{n'} & \dfrac{n}{n'} \end{bmatrix}. \tag{3.34}$$

The reason for the subscript HH' will soon become apparent. We still have not answered the question of how we can convert any system matrix into a thin lens representation without changing the optical system itself. We make a guess that adding some distances to the system might do the trick and they are shown as d and d' in figure 3.10.

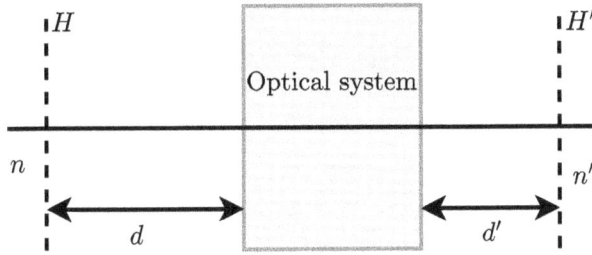

Figure 3.10. Introducing principal planes in an optical system.

The matrix of this system, taking these distances into account, can be obtained from

$$M_{HH'} = \begin{bmatrix} 1 & d' \\ 0 & 1 \end{bmatrix} \begin{bmatrix} A & B \\ C & D \end{bmatrix} \begin{bmatrix} 1 & -d \\ 0 & 1 \end{bmatrix}. \tag{3.35}$$

Carrying out the multiplication and explicitly writing out the individual equations from equation (3.35) gives us

$$A + Cd' = 1 \tag{3.36}$$

$$Ad + + B + Cdd' + Dd' = 1 \tag{3.37}$$

$$C = \frac{-P}{n'} \tag{3.38}$$

$$Cd + D = \frac{n}{n'}. \tag{3.39}$$

The unknowns of this system are its power P and the distances d and d'. We have four equations and three unknowns, and therefore, we can easily arrive at their values, namely:

$$P = -n'C \tag{3.40}$$

$$d = \frac{\frac{n}{n'} - D}{C} \tag{3.41}$$

$$d' = \frac{1 - A}{C}. \tag{3.42}$$

Sign convention. A positive sign for d and d' means the planes lie outside the system. By outside, we mean that the first plane H at distance d would lie to the left of the vertex and the second plane H' would lie to the right of the vertex.

Substituting equations (3.40)–(3.42) into equation (3.37) yields $AD - BC = n/n'$, which is an expected result. The planes H and H' are called the principal planes of the system. We need to confirm that they truly will allow us to treat any system as a

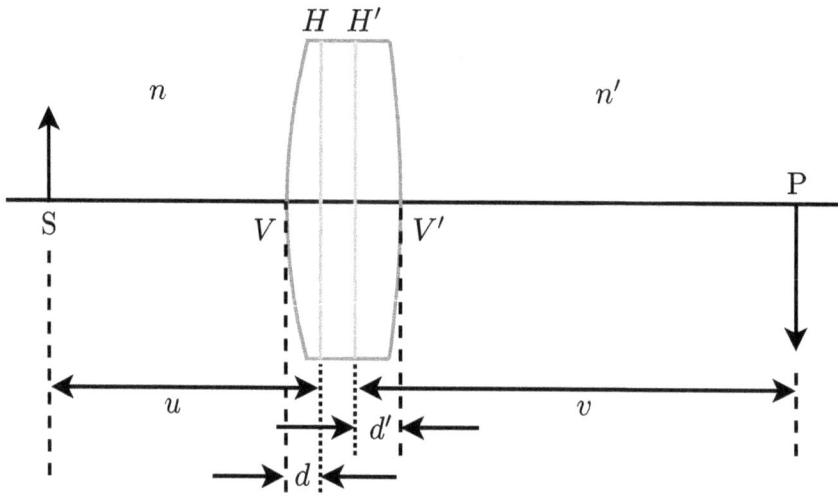

Figure 3.11. Principal planes with respect to the vertices of the optical system.

thin lens. To do this, figure 3.10 is redrawn as shown in figure 3.11, where distances are measured from the principal planes. In this figure, the optical system is shown as a thick lens. However, the system could consist of a number of lenses, in which case the principal planes could lie anywhere inside or outside it.

Example 3.5. Imagine a system comprising two thin lenses of focal lengths 15 cm and 20 cm, with a gap of 10 cm between them. The $ABCD$ matrix is

$$\begin{bmatrix} 0.3333 & 0 \\ -0.0833 & 0.5 \end{bmatrix}.$$

Using equations (3.40)–(3.42), we calculate P or the focal length. It has the value $f = 12.005$ cm. The principal plane locations are $d = -6$ cm and $d = -8$ cm. A schematic of the system along with these planes is shown in figure 3.12. The negative signs imply that they both lie *within* the optical system. In fact, in this example, H' occurs to the left of H.

Example 3.6. If the principal planes transform a system into one that can be analysed like a thin lens then the following equations, based on figure 3.11, should be true:

$$\begin{bmatrix} A & 0 \\ C & D \end{bmatrix} = \begin{bmatrix} 1 & v \\ 0 & 1 \end{bmatrix} \begin{bmatrix} 1 & 0 \\ \dfrac{-P}{n'} & \dfrac{n}{n'} \end{bmatrix} \begin{bmatrix} 1 & u \\ 0 & 1 \end{bmatrix}. \tag{3.43}$$

The lhs of the equation is valid for any conjugate imaging system.

$$f_1 = 15 \text{ cm} \qquad\qquad f_2 = 20 \text{ cm}$$

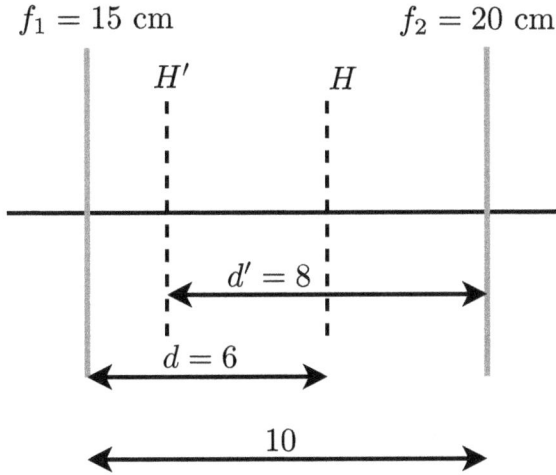

Figure 3.12. Example showing the location of H and H' within the optical system.

$$= \begin{bmatrix} 1 - \dfrac{vP}{n'} & u - \dfrac{uvP}{n'} + \dfrac{vn}{n'} \\[2mm] \dfrac{-P}{n'} & \dfrac{-uP}{n'} + \dfrac{n}{n'} \end{bmatrix}. \tag{3.44}$$

By equating the B coefficient to 0, we obtain

$$\frac{n}{u} + \frac{n'}{v} = P, \tag{3.45}$$

which is exactly the form of the Gaussian imaging equation given by equation (2.12). The difference is that the latter equation was derived with distances measured from the vertex of a thin lens, whereas in this case the distances are measured from the principal planes.

If the object u is at ∞, then from equation (3.45) v should represent the back focal length (BFL) of the system given by BFL $= n'/P$. Similarly, if the image is at ∞, then the front focal length (FFL) is given by FFL $= \frac{n}{P}$. Both these terms are measured from their respective principal planes and they are not equal if the object and image refractive indices are not identical.

3.4 Physical meaning of matrix elements

3.4.1 The A coefficient

If the conjugate system imaging matrix (given in the lhs of equation (3.43)) is used in equation (3.8), the A coefficient is equal to

$$A = 1 - \frac{vP}{n'}, \tag{3.46}$$

3-15

but this is nothing other than the transverse or linear magnification M of a system. Therefore,

$$M = A = \frac{y_2}{y_1}.$$
(3.47)

Replacing P with its representation from equation (3.45),

$$M = \frac{-nv}{n'u}.$$
(3.48)

3.4.2 The B and C coefficients

We have already seen that $B = 0$ means that the system is an imaging system and C relates to the power of the system.

3.4.3 The D coefficient

This coefficient relates to the angular magnification M_α of the system and is of relevance in optical instruments with eyepieces, such as microscopes, binoculars and telescopes. Angular magnification is defined as the ratio of the size of the image on the retina, as seen through the instrument, to the image size seen by the naked eye when the object is at the near viewing point. For humans, this point is approximately 25 cm away from the eye and it is the closest an object can be to the eye and still be in focus. The size of the image formed on the retina is related to the angle subtended by the object. This angle is called the visual angle. It should be noted that when looking at objects with the eye, a lot of what we see is determined by the brain and not just the optics of the eye. The visual angle plays a strong role in the perceived size of an object. Two different scenarios are shown in figures 3.13(a) and (b).

One might wonder why angular magnification is needed, when there already exists a measure of linear magnification. Let us say that you are looking at the Moon with your eyes, the visual angle θ_1 is approximately 0.5°, given that the diameter and its distance from earth are approximately 3474 km and 383 000 km, respectively. However, if you were to use a pair of binoculars, the virtual image formed would be much closer causing the visual angle to be larger. A pair of binoculars with a magnification of 10× would mean that the visual angle subtended by the Moon through the binoculars is now 5°. Just to be clear: this means that the image *appears* 10× bigger on your retina than when looking at the Moon directly. However, the image is not actually 10 × bigger, which would be the case if the transverse magnification was 10.

We now return to relating the D coefficient of a conjugate imaging system, given by equation (3.44), to the angular magnification:

$$M_\alpha = \frac{-uP}{n'} + \frac{n}{n'}.$$
(3.49)

(a)

(b)

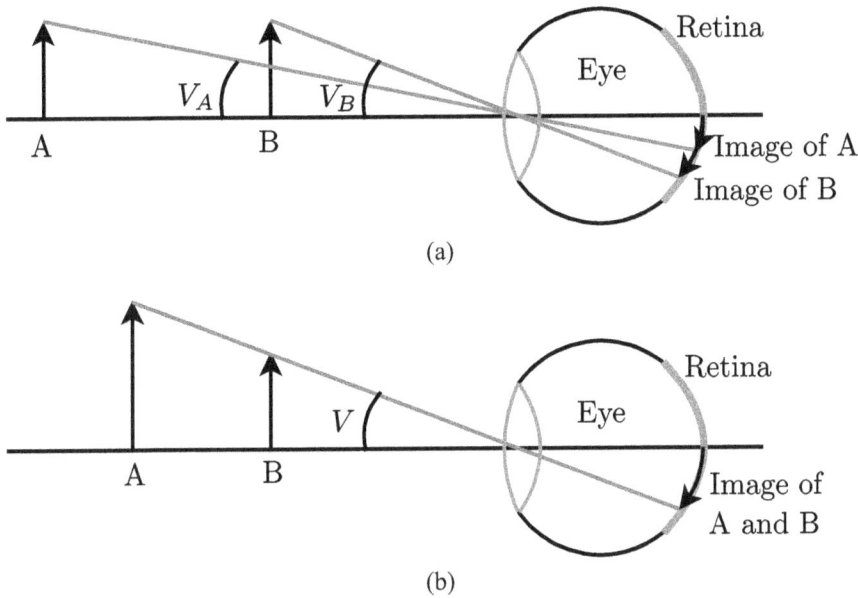

Figure 3.13. Perceived image size for (a) two same-sized objects with different visual angles and (b) two differently sized object with the same visual angle.

Once again, replacing P from equation (3.45), gives us

$$M_\alpha = \frac{-u}{v}. \tag{3.50}$$

The matrix for a conjugate imaging system can finally be written as

$$= \begin{bmatrix} M & 0 \\ \dfrac{-P}{n'} & M_\alpha \end{bmatrix}. \tag{3.51}$$

Calculating the determinant provides a relationship between the transverse and angular magnifications of such a system since $MM_\alpha = n/n'$.

3.5 Cardinal points

It has been shown how the matrix method can be applied to systems with multiple components in them. A pair of planes, called the principal planes, were defined from which the image and object distances could be defined. But how do we define the focal length of such a system?

To answer these questions, we use a set of points to translate the ideas of thin lenses to the general case of multi-element systems. The points useful for optical systems are called the cardinal points. Specifically, they are the:

- Principal points.
- Focal points.
- Nodal points.

3.5.1 Principal points

We have already defined the hypothetical principal planes. The points where they cross the optical axis are called the principal points of the system.

3.5.2 Focal points

When working with thin lenses, all distances could be measured from the (single) vertex of the lens. For example, the focal length of such a lens was the distance from the vertex to the point a collimated incident beam was focused to on the axis, as seen in figure 3.14.

Unsurprisingly, for multi-element or thick lenses, the focal length is defined with respect to the principal planes. Unlike a thin lens, however, these kinds of systems have three different measures of focal length; namely the front, back and effective focal lengths or FFL, BFL and EFL, respectively, as shown in figure 3.15.

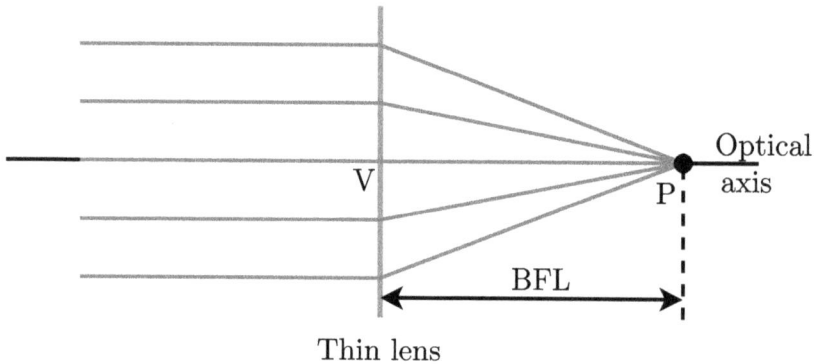

Figure 3.14. Back focal point of a lens.

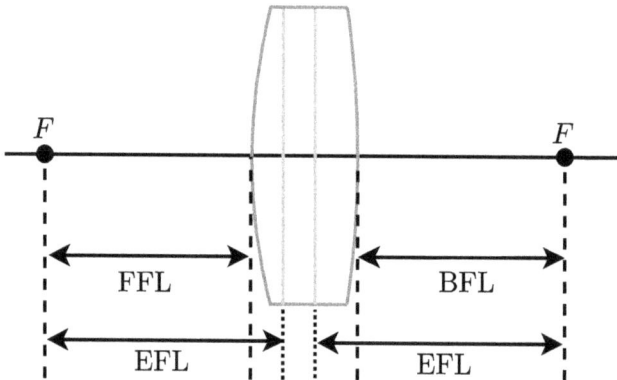

Figure 3.15. Back, front and effective focal lengths of a system.

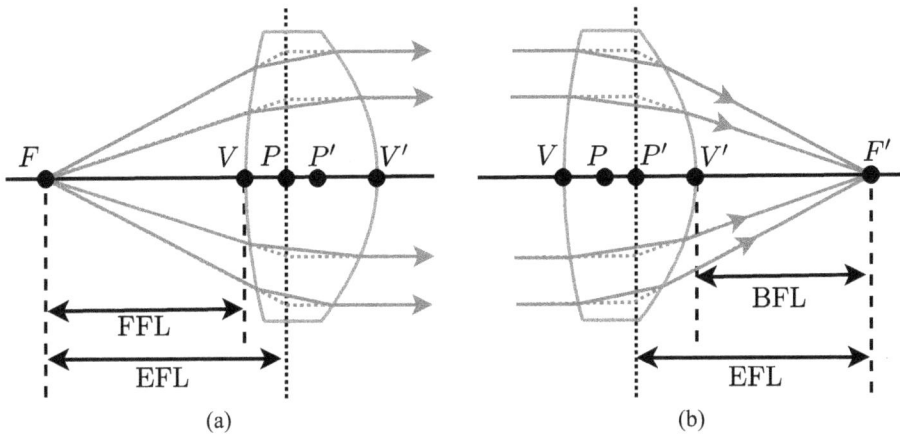

Figure 3.16. Graphically arriving at the principal planes (a) first principal plane and (b) second plane.

The front and back focal lengths of a system are measured from vertices V_1 and V_2, respectively. If the system consists of more than one element, they would be measured from the vertices of the first and last element. It should be noted that a system might not have the same FFL and BFL. However, there is only one EFL, as that is associated with the power of the system, which is independent of the direction that light travels. EFL is always measured from the principal points.

Another way of thinking about or visualising the principal planes is shown in figure 3.16. Rays leaving the focal point on the left of the lens will emerge parallel. Similarly, an incident parallel beam of light will be focused on the focal point after the lens. The principal planes are formed at the intersection of these two. The locations of the planes vary for different lenses and may not even lie within the lens at all!

3.5.3 Nodal points

In a thin lens, a ray through the centre would travel through undeviated. In other words, the emerging ray has the same angle with respect to the optical axis, as the incident ray. The nodal points help define similar rays, that is rays that emerge from the optical system with the same angle they had when incident on the system. Rays travelling towards one nodal point, will appear to have travelled from the other. In figure 3.17, ray AA″ travels to nodal point N and emerges as the parallel ray Bessel beam (BB′) which appears to come from N′.

Point to ponder: Nodal rays are parallel even if the refractive indices on either side of the lens are not identical. How is this possible?

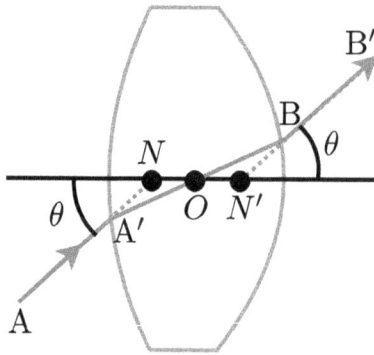

Figure 3.17. Nodal points of a system.

3.6 Bridging the gap between theory and design tools

All optical design tools will generate a report that gives details about the EFL, BFL, locations of the principal planes, etc. These tools also have easy options to ensure that surfaces that represent physical ones are drawn and visible to the user. This is not the case however, for the principal planes. A few extra steps, discussed in appendix A.2, are required to make them visible. This is also tested in problem 6.

3.7 Problems

1. Arrive at the system matrix $\begin{bmatrix} A & B \\ C & D \end{bmatrix}$ for the systems shown in figure 3.18. For each of the figures shown, is there anything special about the matrix components?

2. Two thin lenses, one convex of focal length 0.2 m and the other concave of focal length 0.1 m, are placed on a common axis 0.08 m apart. An object with height 0.01 m is placed at a distance of 0.4 m from the convex lens. Calculate the position and size of the image using the matrix method.

3. Locate the principal planes of the optical system of the previous problem.

4. Using the concept of the Lagrange Invariant, determine whether the image can be as shown in figure 3.19.

5. The first lens of a telescopic system has a diameter of 15 cm. The system is to be used in the visible range. The detector has a diameter of 1 mm and can collect light in a cone up to $\pm 20°$ from normal. What is the maximum FoV of this instrument?

6. *Locating principal planes using OSLO or Zemax*
   What are the back and effective focal lengths of a system (working at a wavelength of 550 nm) that consists of two thick lenses, both made of N-BK7, 10 mm apart? The first lens has radii of curvature 89.672 and −45 mm, respectively, whereas the second lens (of thickness 4 mm) has radii of curvature −70 and 91.645 mm. Draw the virtual principal planes on the optical layout.

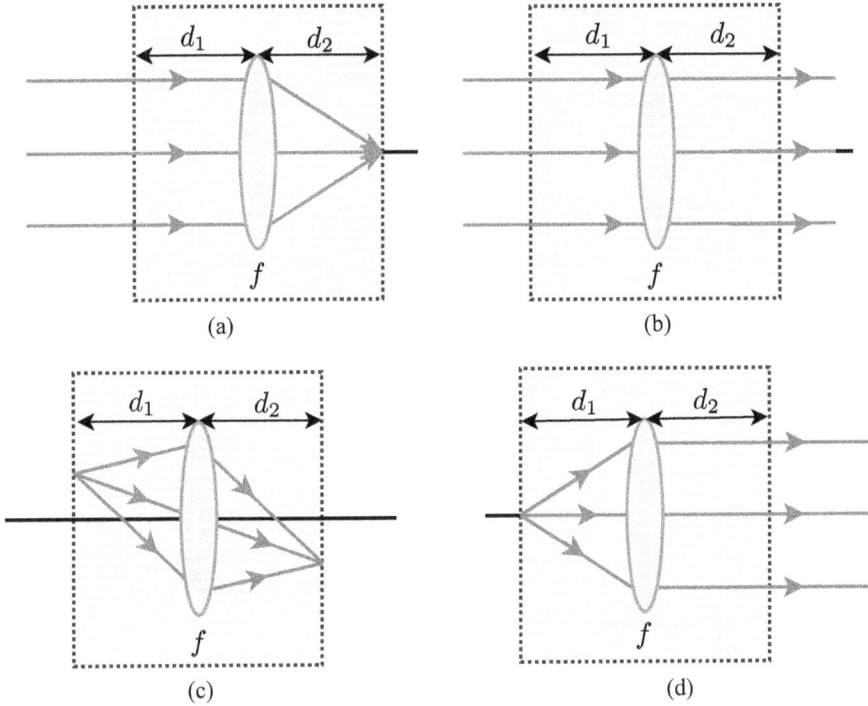

Figure 3.18. Different optical systems.

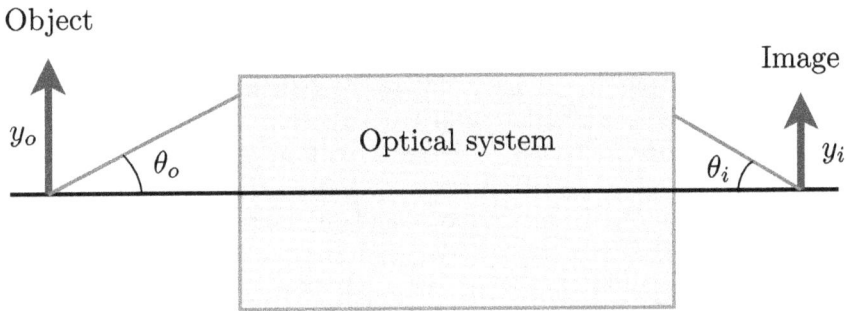

Figure 3.19. Using LI to ascertain whether the output of an optical system is realistic or not.

References

[1] Steinich T and Blahnik V 2012 Optical design of camera optics for mobile phones *Adv. Opt. Techn.* **1** 51–8
[2] Sasián J 2019 *Introduction to Lens Design* (Cambridge: Cambridge University Press)
[3] Saleh B E A and Teich M C 2019 *Fundamentals of Photonics* (Wiley Series in Pure and Applied Optics) (New York: Wiley)
[4] Greivenkamp E J 2004 *Field Guide to Geometrical Optics* (Bellingham, WA: SPIE Press)
[5] Oslo Optics Reference *Lambdares.com* https://lambdaresfiles.com/wp-content/uploads/support/oslo/oslo_releases/OSLOOpticsReference.pdf. (Accessed: 29 March 2023)

IOP Publishing

Introduction to Ray, Wave, and Beam Optics with Applications

Shanti Bhattacharya

Chapter 4

Aberrations

Ideally, a point object should create a point image. We have seen that even the most perfect practical systems are always diffraction-limited, resulting in a spot of finite width. This is due to the finite nature of the apertures involved. In addition, real systems are prone to aberrations, due to a variety of causes. For instance, our assumption that all rays passing through the system are paraxial may not be valid. The main effect of aberrations is that all rays from one object point do not reach one image point, as seen in figure 4.1.

A large part of optical design lies in designing optical systems with minimum aberrations. In order to do so, one needs to understand what causes the different kinds of aberrations and, therefore, optimise the system to eliminate or minimise them in the best possible way. For our discussions, unless otherwise stated, we look at surfaces that possess circular symmetry with respect to the optical axis. Aberrations will be quantified by tracing rays from the object to image space. While any ray can be traced, the meridional and sagittal rays are particularly useful. These rays were defined in section 2.7. Meridional rays lie in the $y-z$ plane and have angle θ equal to 0 or π radians, as in figure 4.1. Sagittal rays, on the other hand, lie in an orthogonal plane with angles $\theta = \pi/2$ or $3\pi/2$ radians. When coming from an off-axis point, the plane they lie in will not contain the optical axis.

4.1 Means of quantifying aberrations

Lenses without aberration form an image at a point determined by paraxial equations. In systems with aberrations or in the presence of non-paraxial rays, how do we quantify the aberration? There are many different ways of doing this, as we will see in this chapter. A simple method might be to measure the amount by which the rays miss the paraxial image point in the transverse image plane. Or we could measure the distances the rays cross the optical axis with respect to where they ought to cross the axis. This is more of a longitudinal measure. Before we go into the details of quantifying these ray aberrations, we introduce the idea of quantification

doi:10.1088/978-0-7503-5497-4ch4

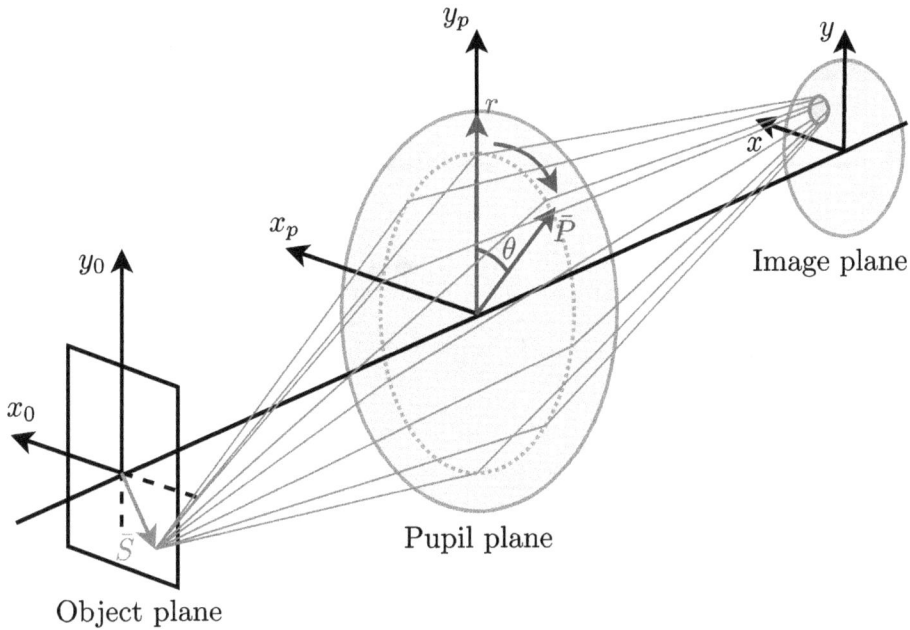

Figure 4.1. Rays travelling from the object plane through the pupil that represents the optical system to the image plane.

through waves. It should not be surprising that waves can give us the same information that rays do, as rays are nothing but normals to the waves. Wave aberrations are important because they can actually be measured (using interferometric techniques), which is not the case for ray aberrations.

In object space, we can consider wavefronts to be perfect concentric spheres centred about the object point. However, in an aberrated system, the wavefronts in image space are no longer perfect spheres. This deviation from a perfect reference sphere is a measure of the wave aberration. Clearly, the radius and centre of the reference sphere must be chosen wisely. Usually, it is determined by locating the reference sphere centre at the ideal image point. The reference sphere is drawn about this centre such that the intersection point of the optical axis and the chief ray lies on its surface. This intersection point basically denotes the location of the exit pupil (see section 2.7). This is schematically demonstrated in figure 4.2, which shows a wave travelling through an optical system with aberrations.

The ideal wavefront in image space (that would exist if the optical system were aberration-free) is shown in green and the aberrated one is in red. The wave aberration W can be defined in terms of a path length difference between the ray from the reference sphere (i.e. un-aberrated wavefront) to the aberrated wavefront. Or in other words,

$$W = \mathrm{SQ}_2 - \mathrm{SQ}_1 = \mathrm{SO} - \mathrm{SQ}_1. \tag{4.1}$$

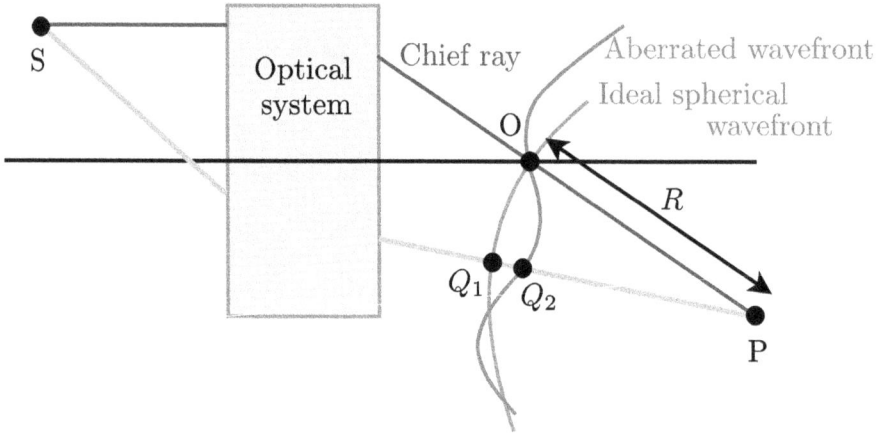

Figure 4.2. Defining a reference sphere to enable calculation of wave aberration.

The rhs of the equation is true because points O and Q_2 in figure 4.2 lie on the same aberrated wavefront, which means they have the same optical path length.

In general, the wave aberration can be described in terms of the object point coordinates, x_o, y_o and the pupil coordinates x_p and y_p or polar coordinates r and θ, given in figure 4.1. So W from equation (4.1) can be written as $W(x_o, y_o, x_p, y_p)$. It is important to clarify what is meant by pupil here, as some texts deriving the wave aberrations will do so in terms of the coordinates at the entrance pupil and others will use the coordinates at the exit pupil. Keep in mind that the entrance and exit pupils of a system are not independent of each other and are determined by the aperture stop of the system. Therefore, the location of the image points can be determined in terms of either the entrance or exit pupil ray coordinates but do not require both. A ray travelling from the object point of interest through the chosen pupil coordinates will finally intersect the image plane at (x, y). We note that because of the rotational symmetry of the system, the wave aberration will not change as the ray is rotated. We can take this into account in our wave aberration by using parameters which are functions of the object and pupil coordinates, instead of using them directly. We calculate three parameters in terms of x_o, y_o, x_p and y_p. They are

- The square of the object vector length $\bar{S} \cdot \bar{S} = x_o^2 + y_o^2$.
- The square of the pupil vector length $\bar{P} \cdot \bar{P} = x_p^2 + y_p^2$.
- The scalar product of the object vector and the pupil vector $\bar{S} \cdot \bar{P} = x_p \cdot x_o + y_p \cdot y_o$.

Again, because of the circular symmetry, the object can be considered a linear vertical line with its base on the optical axis, or in other words, we choose $x_o = 0$. In this case, we can think of the coordinate in the y-axis as the height of the object, so $y_o = h$. The wavefront then becomes a function of

$$W = W\left(h^2, \ x_p^2 + y_p^2, \ y_p \cdot y_p h\right). \tag{4.2}$$

This can be expanded in terms of all the possible power series terms as

$$
\begin{aligned}
W = & A_1\left(x_p^2 + y_p^2\right) + A_2 h y_p + A_3 h^3 \\
& + B_1\left(x_p^2 + y_p^2\right)^2 + B_2 h y_p\left(x_p^2 + y_p^2\right) + B_3 h^2 y_p^2 \\
& + B_4 h^2\left(x_p^2 + y_p^2\right) + B_5 h^3 y_p + B_6 h^4 \\
& + C_1\left(x_p^2 + y_p^2\right)^3 + C_2 h y_p\left(x_p^2 + y_p^2\right)^2 + C_3 h^2 y_p^2\left(x_p^2 + y_p^2\right) + \\
& \quad C_4 h^2\left(x_p^2 + y_p^2\right)^2 + C_5 h^3 y_p\left(x_p^2 + y_p^2\right) + C_6 h^3 y_p^3 \\
& + C_7 h^4 y_p^2 + C_8 h^4\left(x_p^2 + y_p^2\right) + C_9 h^5 y_p + C_{10} h^6 + \cdots .
\end{aligned}
\tag{4.3}
$$

Because of the way the wave aberration is defined in equation (4.1), it goes to zero when $x_p = 0$, $y_p = 0$. Therefore, any coefficient that does not depend on either x_p or y_p must also be 0. This means coefficients A_3, B_6, etc, are zero. (These terms are relevant when taking pupil aberrations into account but can be ignored in this discussion.) Neglecting these terms and using polar coordinates reduces the wave aberration to

$$
\begin{aligned}
W = & A_1 r^2 + A_2 h r \cos\theta \\
& + B_1 r^4 + B_2 h r^3 \cos\theta + B_3 h^2 r^2 \cos^2\theta \\
& + B_4 h^2 r^2 + B_5 h^3 r \cos\theta \\
& + C_1 r^6 + C_2 h r^5 \cos\theta + C_3 h^2 r^4 \cos^2\theta + \\
& \quad C_4 h^2 r^4 + C_5 h^3 r^3 \cos\theta + C_6 h^3 r^3 \cos^3\theta \\
& + C_7 h^4 r^2 \cos^2\theta + C_8 h^4 r^2 + C_9 h^5 r \cos\theta + \cdots .
\end{aligned}
\tag{4.4}
$$

The transverse aberrations x and y, measured at the image plane, can be calculated [1] by differentiating the wave aberration equation (4.3) with respect to x_p and y_p using

$$
x = \frac{R}{n'}\frac{\partial W}{\partial x_p}
\tag{4.5}
$$

$$
y = \frac{R}{n'}\frac{\partial W}{\partial y_p}.
\tag{4.6}
$$

With the help of some trigonometric relationships, the transverse aberrations finally become

$$
\begin{aligned}
x = \frac{R}{n'}(& 2A_1 r \sin\theta + \\
& + 4B_1 r^3 \sin\theta + B_2 h r^2 \sin 2\theta + \\
& + 2B_4 h^2 r \sin\theta + \cdots)
\end{aligned}
\tag{4.7}
$$

and

$$
\begin{aligned}
y = \frac{R}{n'}(&2A_1 r \cos\theta + A_2 h \\
&+ 4B_1 r^3 \cos\theta + B_2 h r^2 (2 + \cos 2\theta) \\
&+ 2(B_3 + B_4)rh^2 \cos\theta + B_5 h^3 + \cdots).
\end{aligned}
\tag{4.8}
$$

Already, we can obtain some information from these equations. The order of the aberrations are nothing other than the sum of the powers of the object (height) and pupil variables. For example, the powers of the terms associated with the A coefficients is unity, whereas it is 3 for the B coefficient terms and so on. There are no even sums, as this is a system with radial symmetry. The way this power is split between both variables gives us insights into the types of aberrations. There is a slight distinction between the A terms and all the other terms. The former are not really aberrations but are associated with paraxial imagery. A_1 provides a measure of the distance from the paraxial focus to the current image plane. It can be considered as a measure of defocus of the system and, in principle, is easy to correct (with an image plane shift). On the other hand, A_2 is the magnification y/h of the system. The other terms are the transverse aberrations, which are a measure of how much the ray deviates from the ideal image point. The B terms are referred to as the primary or third-order aberration coefficients [2]. They are also commonly called the Seidel aberration coefficients. The C terms are the secondary or fifth-order aberration coefficients. Optical design tools will offer a multitude of different ways of displaying aberrations, including tabulated values or graphical outputs. The aberrations may be presented as geometric or wave aberrations. In the following chapters, we will refer to the Seidel aberrations as given by equations (4.7) and (4.8). However, many textbooks and optical design tools also talk about aberrations in terms of Seidel sums, usually denoted by S_I, S_II, S_III, S_IV and S_V, which can be related to the Seidel coefficients. The Seidel sums are a form of unconverted aberrations. Using the terms in equations (4.7) and (4.8), these numbers can be converted to transverse or angular aberrations [1]. The former is normally used for focal systems and the latter for afocal ones.

They consist of expressions that are defined at each surface of an optical system in terms of the chief and marginal ray heights and angles, the Lagrange invariant of the system, the refractive indices before and after the surface, etc. One of the main benefits of working with the sums is that they can be calculated at each surface of a system and added up to get the total Seidel sum for the system. This gives the benefit of understanding which element contributes the most to a particular aberration and also what needs to be added to the system to cancel out that aberration. By using appropriate equations, this total can be converted to the transverse or wave aberration. At this level, however, it is not necessary to go into the finer details of the Seidel sums. Interested readers are invited to read [1] or [3] to understand them better. We continue this discussion in terms of the Seidel aberrations.

4.2 Monochromatic aberrations

These are monochromatic aberrations denoted by the B terms of equations (4.7) and (4.8). Specifically, B_1 is associated with spherical aberration, B_2 with coma, B_3 with astigmatism, B_4 with Petzval curvature and finally B_5 with distortion. At this point, we look at each of these aberrations individually. In a real system, however, more than one aberration is likely to occur simultaneously and the resulting image will be the result of all the aberrations present.

4.2.1 Spherical aberration

We already know that the spherical shape is not the ideal shape for imaging. Spherical aberrations are a direct outcome of this. More specifically, we can define the aberration as one that arises because different radial zones of a spherical lens have different focal points. The effect is more visible the larger the diameter of the lens or the greater the deviation from the paraxial approximation is. Spherical aberration occurs for rays that come from an on-axis object point. This fact is implied in the B_1 terms, as the object height does not play a role in either equation (4.7) or equation (4.8). The presence of r, however, means that the value of the aberration depends on the radial position the ray comes from. This is one aberration which can be studied without the presence of other aberrations. The typical aberration as seen by a convex lens, known as undercorrected spherical aberration, is shown in figure 4.3.

From the figure, one can see that for every ray hitting the top of the lens, there is a ray with an equal but opposite angle incident on the bottom of the lens. This is true for the rays in the sagittal plane as well. This correspondence is reflected in the effect of the aberration, which can be observed as a symmetrical blur around the focused spot. There are three planes of interest in this figure. The rays incident on the optical system, furthermost from the optical axis, come to focus at plane 1. The paraxial rays, on the other hand, come to focus at plane 3. But in between these two planes, is plane 2 where the spot has the smallest diameter. This spot is called the circle of least confusion (CLC). Often it makes more sense to use this plane as the image plane, rather than the paraxial image plane.

In figure 4.3, we see different measures of the aberration, namely:

1. Longitudinal aberrations are measured along the optical axis. Longitudinal spherical aberration (LSA) is a measure of the distance between where the rays from the extreme of the lens cross the optical axis as opposed to where the paraxial image is. Convex lenses have undercorrected or negative spherical aberration because the rays incident at the extreme ends of the lens have a shorter focal length. LSA or error in focal length of a lens of focal length f and diameter D can be estimated using the formula [4]

$$\Delta f \approx \frac{D^2}{8f(n-1)}. \tag{4.9}$$

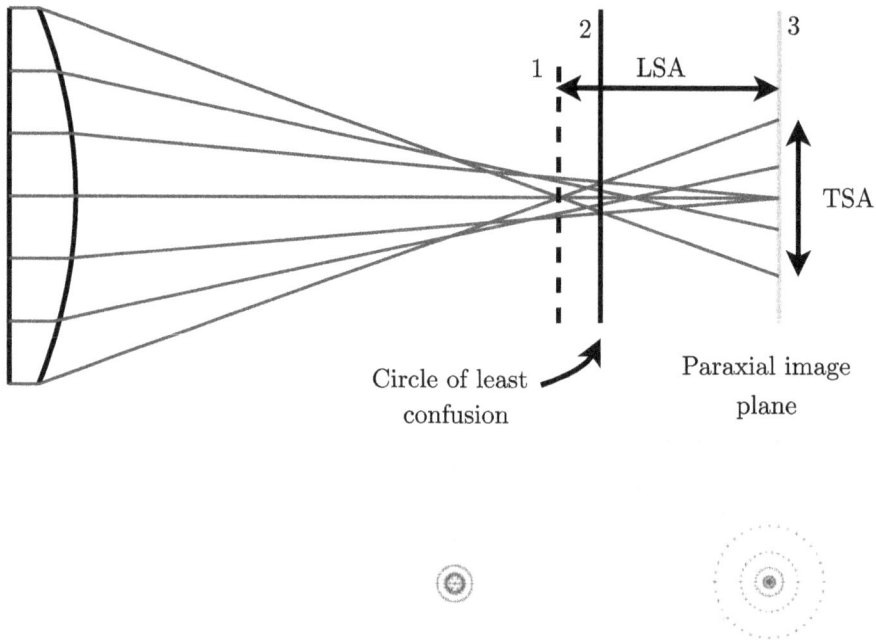

Figure 4.3. Spherical aberration. The spot diagrams are shown below.

2. Transverse spherical aberration (TSA) is measured at the paraxial image plane. The deviation of the rays focused at the plane labelled 1 from the paraxial point at plane 3 gives a measure of this aberration.

The Seidel coefficients are another form of unconverted aberrations [5]. For example, B_1 and B_2 coefficients could be converted to values proportional to the transverse spherical aberration and transverse coma (TCOMA) using TSA $= B_1 r^3$ and TCOMA $= B_1 r^2 h$, respectively. While these numerical measures of an aberration are important, it is sometimes more useful to graphically survey the error with respect to more than just the extreme rays.

4.2.2 Longitudinal aberrations

The longitudinal aberration is measured with respect to ray height. Unlike typical graphs, this one is plotted with the independent variable on the y-axis (rays from greater heights will have a larger y-value) and the error is plotted on the x-axis. Pupil coordinates are usually normalised in optical design software, such that the ray furthermost from the optical axis in the $+y$ direction will have coordinate $+1$, while the axially incident ray will have coordinate 0. For undercorrected LSA, the curve bends towards the negative x-axis, as the error gets more negative for rays further away from the axis. This is shown in figure 4.4(a).

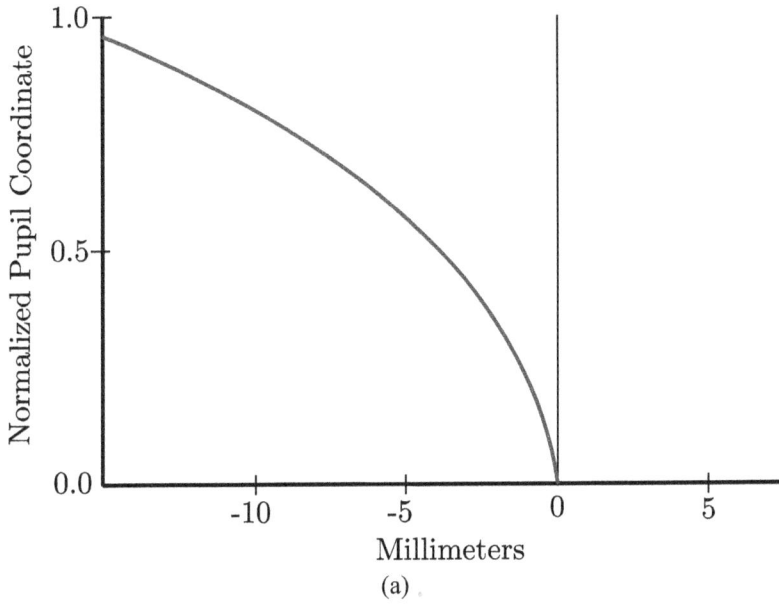

(a)

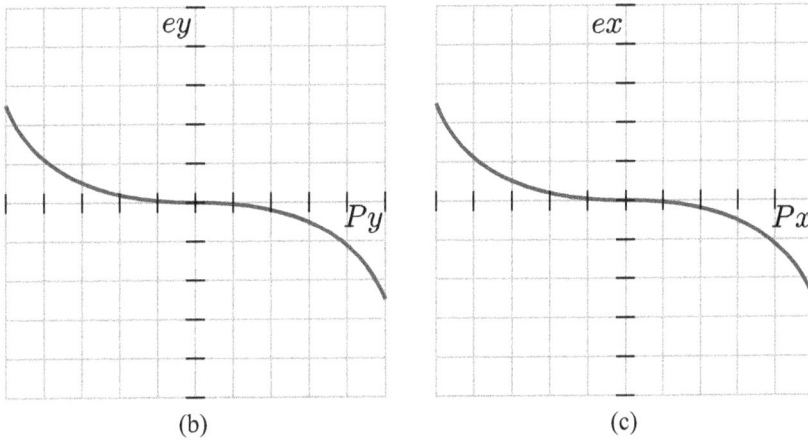

(b)

(c)

Figure 4.4. Graphical representation of (a) longitudinal spherical aberration, (b) transverse spherical aberration for meridional rays and (c) transverse spherical aberration for sagittal rays.

4.2.3 Transverse aberrations

In a system with no aberration, rays crossing any point of the pupil aperture would all arrive at the ideal paraxial image point. In practice, rays crossing the aperture at different coordinates miss the ideal image point (on the image plane) by varying amounts. The transverse aberration plot is also called a *ray intercept curve*. The error is plotted on the y-axis, while the pupil coordinate of that ray is the x-axis coordinate. Often two curves are drawn. Both give the error in the image plane;

one plots the meridional rays along the y-axis (Figure 4.4(b)) and the other the sagittal rays along the x-axis (figure 4.4(c)).

Point to ponder: Can meridional rays become sagittal rays and vice versa, as rays travel through a system?

While some aberrations can be displayed using either transverse or longitudinal aberrations (namely, spherical aberration, coma, astigmatism, field curvature and axial chromatic aberration), field curvature and astigmatism are easier to understand when viewing them as longitudinal aberrations, as will be seen in subsequent sections. The nature of distortion and lateral chromatic aberration, however, mean they can only be displayed as transverse aberrations.

4.2.4 Spot diagrams

Yet another way to visualise aberrations is to look at the intersection of a fan of rays with the image plane. Such diagrams will look like an array of spots. By choosing the fan of rays wisely, a lot of information of the aberration can be obtained. Figure 4.3 shows the spot diagrams from planes 2 and 3. The spot of the circle of least confusion is clearly more tightly packed. The spot diagrams give rise to even better ways of quantifying aberrations. For example, rather than choosing the plane with the smallest diameter spot as the best focal plane, one could calculate the energy distribution of the spot. Let us say the focus spot of a system was measured at two different planes. One had a larger overall diameter but 90% of its energy lay within a smaller circle than the other one. The larger diameter spot might still be a better focus. In particular, if more of the energy lies within the Airy diameter for that system, that would be a better plane to consider, instead of the CLC plane. For asymmetric aberrations, one could use the spot diagram to locate the centre of gravity of the focus spot. Its displacement from the paraxial image position would then be a measure of that particular aberration.

4.2.5 Coma

Unlike spherical aberration, coma arises because of rays from an off-axis object point, which immediately eliminates the possibility of this aberration being symmetrical. This idea plays a role in other aberrations as well, so let us explore it in some more detail. In figure 4.5(a) a set of meridional and sagittal rays, from an off-axis point, are shown travelling through an optical system.

The difference in the way the lens 'sees' the meridional and sagittal rays should be obvious from figures 4.5(b) and (c). One can think of coma as the variation of magnification with aperture. In this example, since the object lies along the y-axis, the sagittal rays from the extreme object point fall on the lens in a symmetric way,

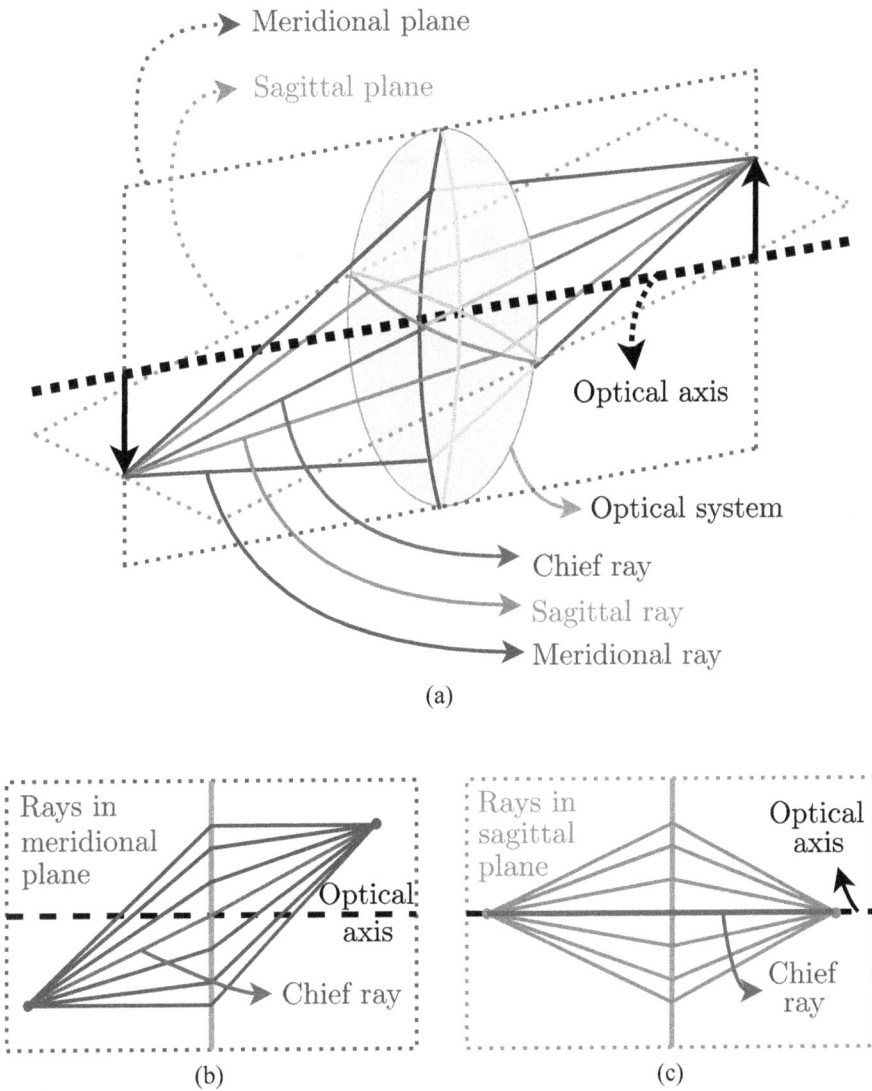

Meridional plane

Sagittal plane

Optical axis

Optical system

Chief ray

Sagittal ray

Meridional ray

(a)

Rays in
meridional
plane

Optical
axis

Chief ray

(b)

Rays in
sagittal
plane

Optical
axis

Chief
ray

(c)

Figure 4.5. (a) Rays from an off-axis point, (b) meridional rays and (c) sagittal rays.

unlike the meridional rays. More specifically, the meridional rays, from an off-axis point, hitting the same radial region at the top and bottom of the lens, will have different angles of incidence, as seen in figure 4.6(a). In figure 4.6(b) we look at two rays with equal (but opposite) angle of incidence, and see that they are incident at two different radial zones.

With a single real lens, it is difficult to display the effects of coma alone on the quality of the image spot, as other aberrations, including spherical aberration, will be present simultaneously. Instead, we can plot the aberration using the B_2 terms in equations (4.7) and (4.8). Unlike the case of spherical aberration, the height h does

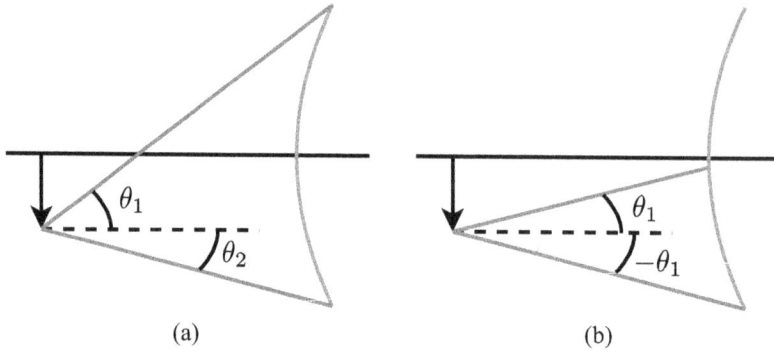

Figure 4.6. Meridional rays from an off-axis point (a) falling on the same radial zone. $|\theta_1| \neq |\theta_2|$ and (b) with equal and opposite angle.

feature in the term. Keeping it constant, i.e. plotting the rays coming from one extreme point, but from all values of r and θ will give rise to a circle in the image plane rather than a spot. Rays from one extreme object point are shown in figure 4.7(a). While the sagittal rays arrive at the same image point, that point is different from the one the meridional rays arrive at. All the rays from one circle (one r value on the optical system) form one circle on the image plane, as shown in figure 4.7(b). However, and this is where coma gets really messy, all the rays from a different r value form a slightly displaced circle as shown in figure 4.7(c). It is in such an image, that locating the centre of gravity can be a good measure of the image location or aberration magnitude.

Coma is studied using the transverse aberration. The curves are shown in figure 4.8. Since the top and bottom meridional rays arrive at the same image point, these curves are symmetric unlike those for spherical aberration.

At the start of the discussion on coma, we said that it is hard to demonstrate it without the effect of other aberrations in a real system. The transverse aberration curves will always include the effect of all the aberrations present. Which aberration is the predominant one can be instantly seen from the shape of the transverse curves.

4.2.6 Astigmatism

Different factors can cause this aberration. If the lens has different radii of curvature in the x- and y-directions, the sagittal and meridional rays (from even an on-axis object point) will come to focus at different planes. The meridional rays lie in the vertical plane (given in red) and the sagittal in the horizontal plane (in blue) in figure 4.9.

An extreme example of this is a cylindrical lens, which has curvature only in one axis. The focus of such a lens is a line, rather than a point. This should be clear from figure 4.9. When the meridional rays come to focus at F_M, the sagittal rays are not focused and therefore form a line at that plane. A similar, but opposite effect, occurs when the sagittal rays come to focus at F_S. However, astigmatism can also occur for

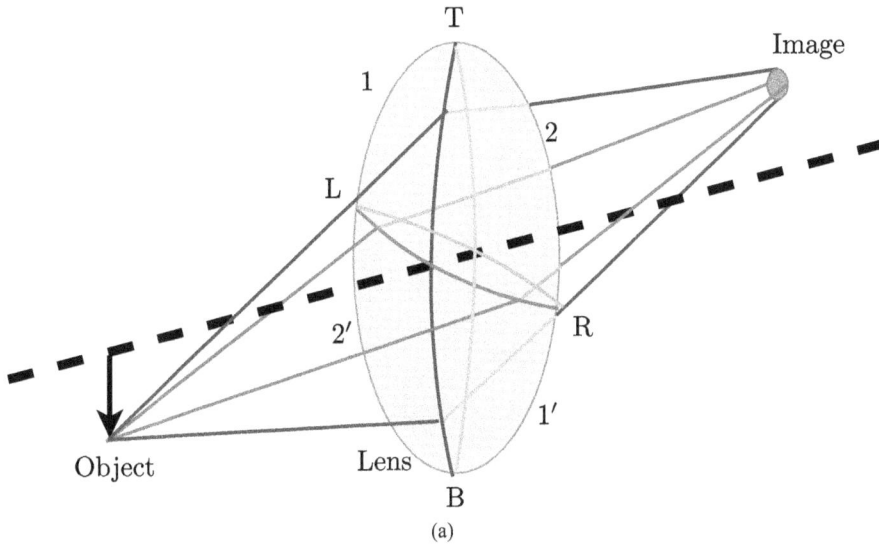

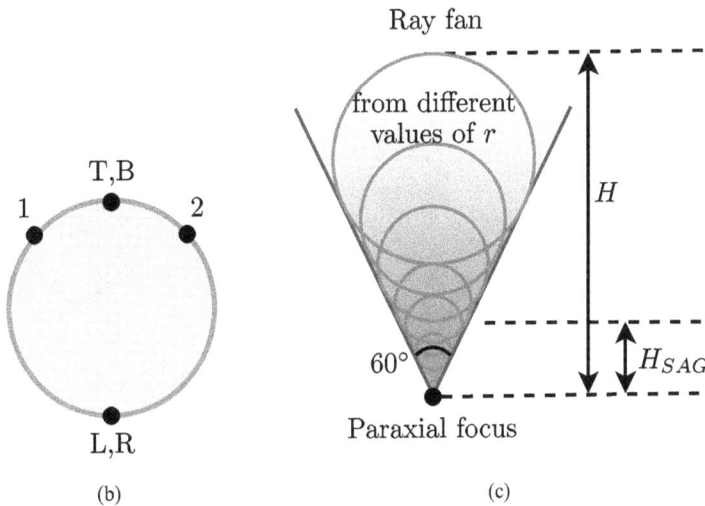

Figure 4.7. Understanding coma. (a) Sagittal and meridional rays with the same radial values forming a circle in the image plane, (b) how the circle is formed and (c) rays from the zone with radius r_1 on the lens form a displaced circle, compared to rays from a different zone with radius r_2.

a spherical lens, when the object rays travel from an off-axis point, as the curvature will appear different in the two orthogonal axes. The effect may seem similar to coma but once again looking at the B_3 terms in equations (4.7) and (4.8) helps clarify the difference. We see that astigmatism depends less on the aperture size (i.e. r) and more on the angle of the incident beam governed by the value of h, since astigmatism has a rh^2 dependence, while coma has a r^3h dependence.

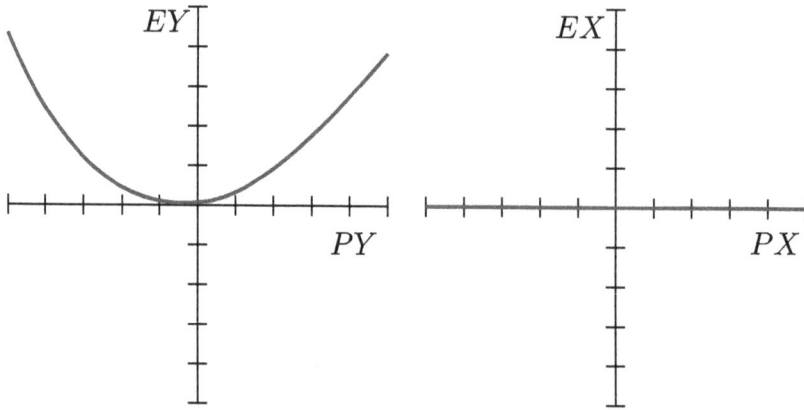

Figure 4.8. Transverse coma curves for meridional and sagittal rays.

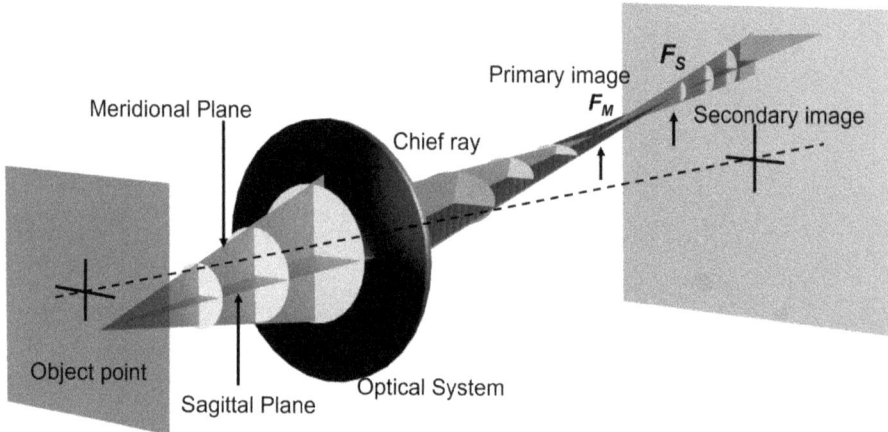

Figure 4.9. Demonstration of astigmatism. Picture credit: RD.

The main difference caused by astigmatism can be seen in the shape of the focus spot. Spherical aberration causes a blur or halo to form around the circular focus spot. Coma creates a comet-like shape smeared out on the paraxial image plane. Astigmatism, on the other hand, creates a rotating line or elliptical focus as one travels along the optical axis. The transverse aberration curves at three planes of relevance are given in figure 4.10.

4.2.7 Field curvature

This aberration arises because the focal plane of a system with curved lenses is actually not planar but curved. This inherent curvature is called the Petzval curvature and can be related to the sum of the product of the refractive indices of the lenses used with their focal lengths. Since meridional and sagittal rays have

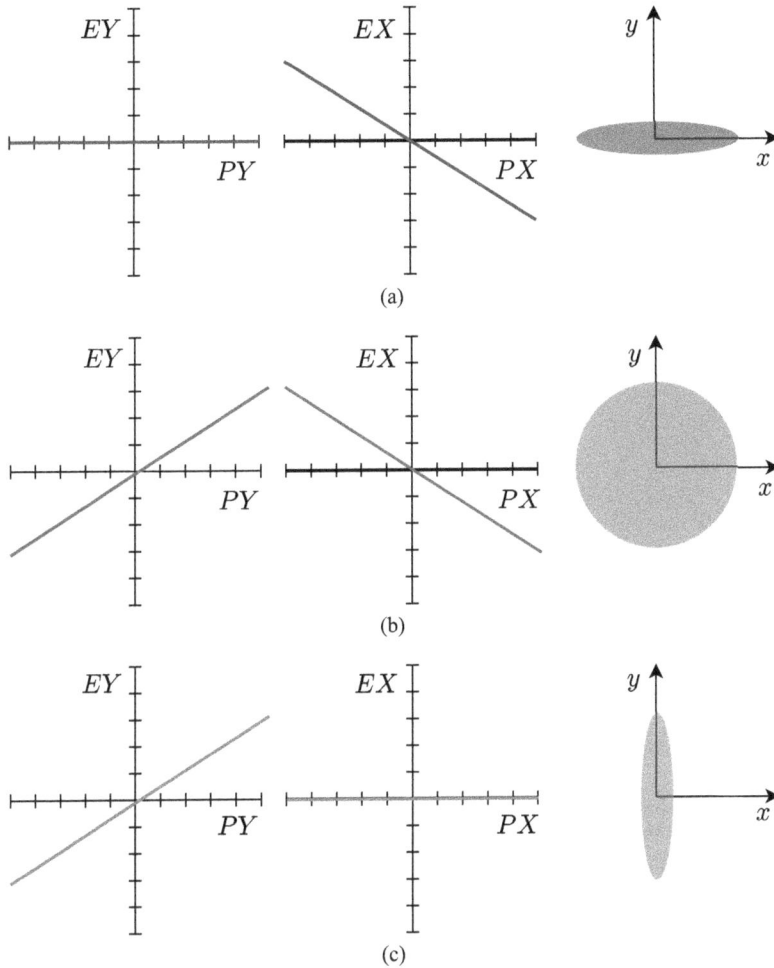

Figure 4.10. Transverse aberration showing aberration at the (a) meridional focus, (b) CLC and (c) sagittal focus.

different foci for off-axis points, the field curvature seen by each set of rays will also be different. This is why any system with astigmatism present will also have field curvature. This aberration is typically quantified or displayed as a longitudinal aberration plot with field curvature curves shown for both tangential (meridional) and sagittal rays. This can sometimes be confusing, as the aberration curves will be labelled meridional and sagittal with no explicit mention of field curvature, leading one to believe that they only represent astigmatism. If there was no astigmatism, the meridional and sagittal curves would both lie on the Petzval surface. If there were no field curvature present, then the image plane would be a flat plane perpendicular to the optical axis. When astigmatism exists in the system, the meridional surface (or curve) lies three times further from the Petzval surface compared to the sagittal surface, as shown in figure 4.11. Most uncorrected imaging systems will have a

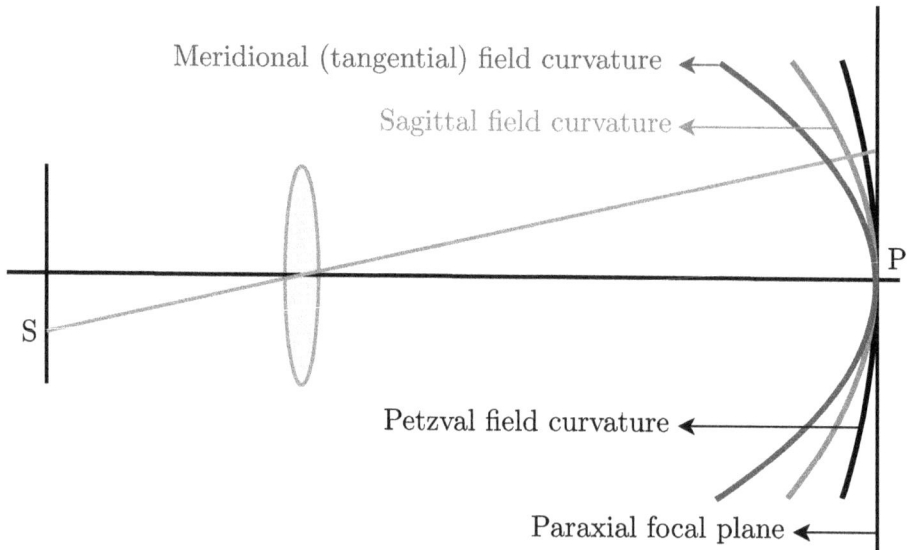

Figure 4.11. Demonstration of uncorrected field curvature.

concave curvature, i.e. the image plane will curve towards the lens system, with both meridional and sagittal curves lying on the same side as the Petzval curve.

> **Point to ponder**: Keeping in mind that this aberration is proportional to the sum of the product of the lenses focal lengths and refractive indices, how could the aberration be corrected and what would be the sum in that case?

4.2.8 Distortion

This aberration is quite different from the others in that a perfectly focused image is formed at the paraxial image plane but as the name suggests, it is distorted. This means points are focused not at their paraxial image point but either closer or further away from the axis. The easiest way to visualise this is to image a square grid, as seen in figure 4.12(a). Distortion is quantified in terms of the ratio of displacement d to the paraxial image height h. d is the radial displacement of the image point from the paraxial image location. It is usually given as a percentage, $100 \times d/h$.

Wide-angle lenses such as fisheye lenses usually suffer from this type of aberration.

4.3 Chromatic aberrations

Chromatic aberration arises because the refractive index of a material is different at different wavelengths. This means that the focal length of a lens is

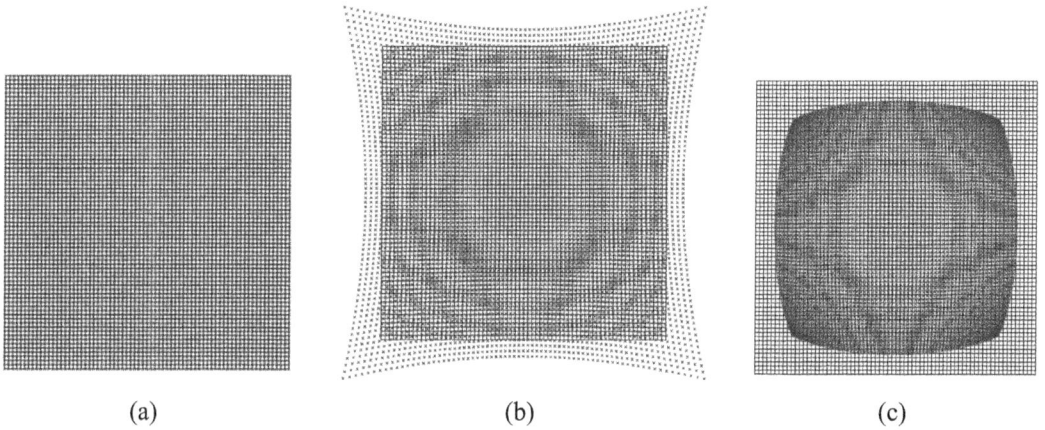

Figure 4.12. Demonstration of different types of distortion. Image of a square grid with (a) no distortion, (b) pincushion distortion and (c) barrel distortion. Picture credit: ST.

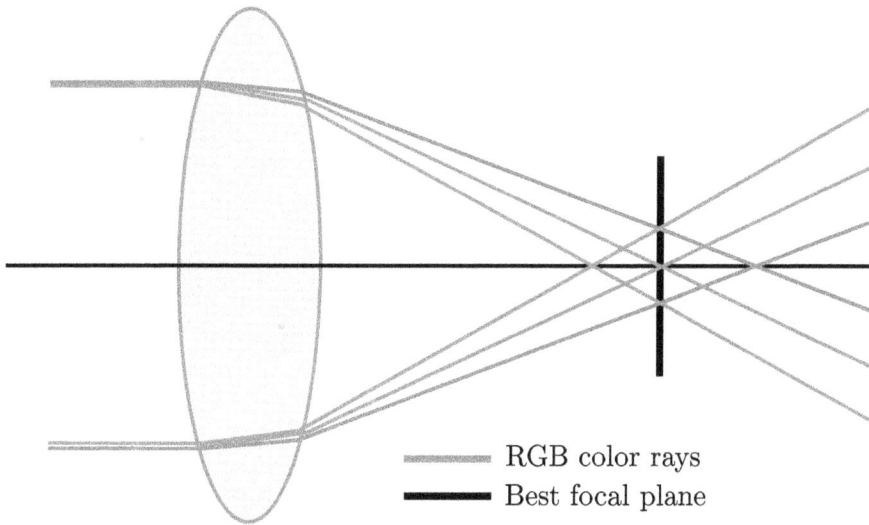

$\cdots\cdots$ RGB color rays
━━━ Best focal plane

Figure 4.13. Longitudinal chromatic aberration.

wavelength-dependent. This aberration can be quantified in two different ways. The longitudinal spread of the focal length along the optical axis is called longitudinal or axial chromatic aberration and is shown in figure 4.13. The central plane of focus would be where the yellow/green light is focused. As can be seen in the figure, this is where the red and blue wavelengths are completely blurred, resulting in a purple haze around the focus spot.

The lateral (or transverse) chromatic aberration shown in figure 4.14 results when an off-axis point has a different magnification for each wavelength. It will appear as a rainbow effect towards the edge of an image.

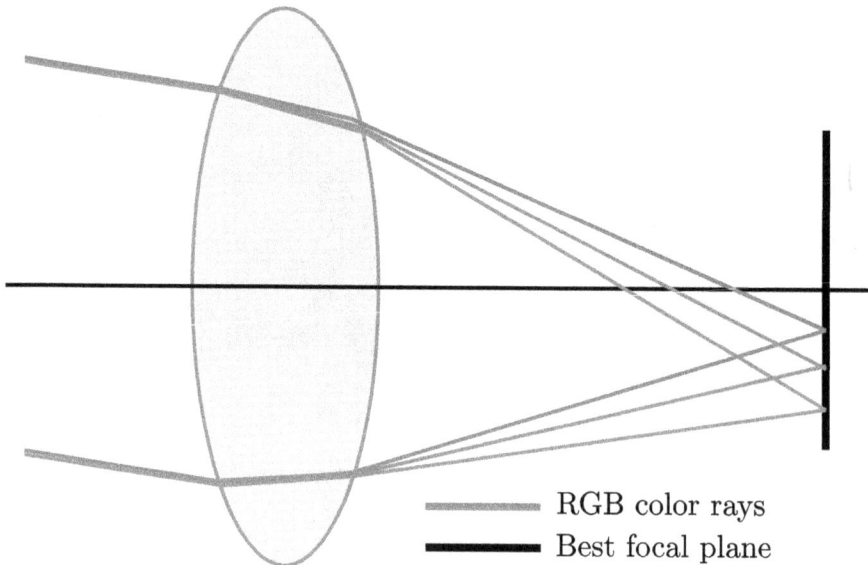

RGB color rays
Best focal plane

Figure 4.14. Lateral chromatic aberration.

4.4 Correcting aberrations

It is important to understand the different factors that can cause aberrations in a system in order to be able to correct for them. There are a number of tools at an optical designers disposal to enable this. They are:

- The power of each element.
- The thickness of an element.
- The shape of each element.
- The stop location.
- The spacing between elements.
- The glass type.

A lens with a small focal length (large power) will tend to have more aberration than one with less power (if the numerical aperture (NA) is the same for both lenses). If there is freedom to choose the power of the lens, it would, therefore, make sense to pick a lens with less power. However, a system will usually have a required power and the aim of aberration correction is to design the system with that power but minimal aberration. In systems with more than one lens, the power of individual lenses could be varied, such that the total power of the system meets the requirement but the combination reduces aberrations. We explore select aberrations in order to explain the methodology behind aberration correction.

4.4.1 Reducing field curvature

Field curvature can be reduced with the help of a Petzval lens, added to an optical system. The lens is a plano-concave lens, which is relatively thin in the middle. It is

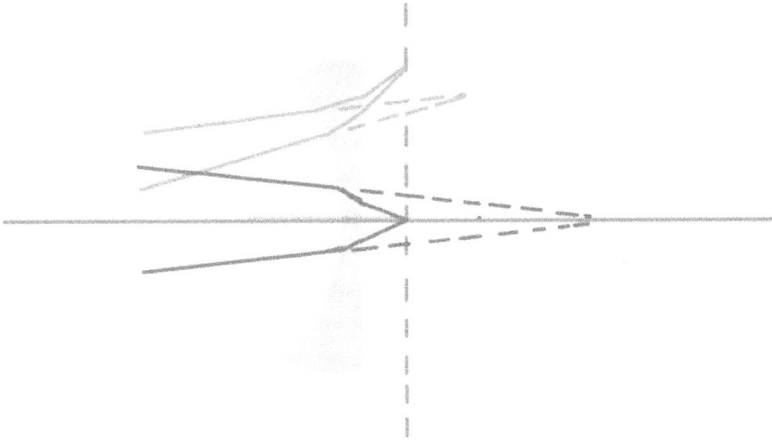

Figure 4.15. The solid lines represent the ray paths taken by rays that travel through the Petzval lens, whereas the dashed rays indicate the paths that would have been taken in the absence of the lens. The Petzval lens brings the green rays to focus on the same image plane as the paraxial (red) rays but there is a lateral displace in a direction perpendicular to the optical axis, which means that there is distortion in the focused image.

placed close to the image plane. There will of course, be a small change in the power of the system but this can be taken into account during the design itself. The trick in the lens is that off-axis rays hitting the extremes of the lens are slowed down more than the rays hitting the centre, as shown in figure 4.15. In other words, rays that would have focused on the (curved) Petzval surface now focus on the paraxial image plane, which helps reduce the field curvature. Often such lenses introduce some distortion but this is considered a small price to pay to avoid defocused light at the extremes of the image. The concepts of both shape and power are used in a Petzval lens to reduce an aberration.

4.4.2 Minimising spherical aberration and coma

In the previous case, a lens was added to an optical system comprising a number of lenses in order to reduce field curvature. Certain aberrations of a single lens can be minimised by changing the shape of the lens [4]. Figure 4.16 plots spherical aberration and coma for lenses of constant power and NA but varying shape.

4.4.3 Correcting axial chromatic aberration

Let us explore the idea of using two lenses made of different materials to correct for chromatic aberration. In general, the resulting power of two lenses with a gap d between them is given by

$$\frac{1}{f} = \frac{1}{f_1} + \frac{1}{f_2} - \frac{d}{f_1 f_2}. \tag{4.10}$$

Using equation (2.11), we can write this in terms of the focal length of each lens. To reduce the terms in the equation, we replace $1/R_1 - 1/R_2$ by the letter ρ_i, which can

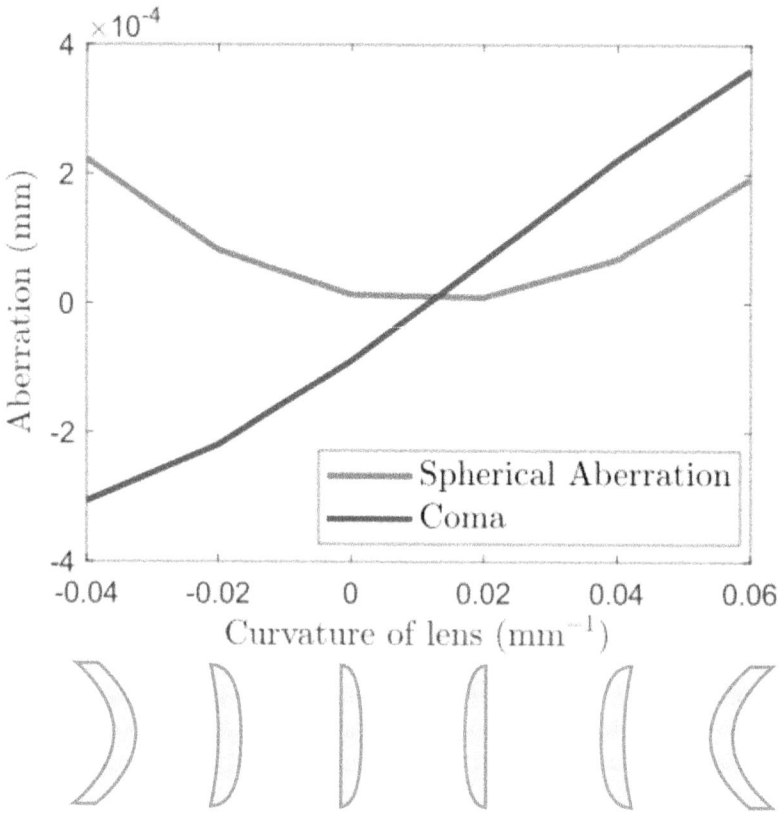

Figure 4.16. Effect of lens shape on aberrations. R_1 (in mm) was varied from $-25, -50, 0, 50, 25, 16.7$, keeping the focal length at 150 mm. Data generated by ST.

be thought of as a factor that represents the shape of the lens i. Equation (4.10) becomes

$$\frac{1}{f} = (n_1 - 1)\rho_1 + (n_2 - 1)\rho_2 - d(n_1 - 1)(n_2 - 1)\rho_1\rho_2. \qquad (4.11)$$

The aim of this exercise was to obtain a system which has the same focal length for two wavelengths, e.g. blue and red. In other words, we require

$$\frac{1}{f_B} = \frac{1}{f_R}. \qquad (4.12)$$

Substituting the relevant terms from equation (4.10), and assuming the lenses are in contact, i.e. $d = 0$ results in

$$\frac{\rho_1}{\rho_2} = \frac{n_{2B} - n_{2R}}{n_{1B} - n_{1R}}. \qquad (4.13)$$

4-19

If yellow light (which has a wavelength between that of blue and red) were sent through the individual lenses, the ratio of their shape factors would be

$$\frac{\rho_1}{\rho_2} = \frac{(n_{2Y} - 1)f_{2Y}}{(n_{1Y} - 1)f_{1Y}}. \tag{4.14}$$

Since the shape of the lens does not change, when different wavelengths travel through it, the two equations (4.13) and (4.14) can be equated, resulting in

$$\frac{f_{2Y}}{f_{1Y}} = -\frac{n_{2B} - n_{2R}}{n_{1B} - n_{1R}} \frac{(n_{1Y} - 1)}{(n_{2Y} - 1)}. \tag{4.15}$$

The Abbe number or V-number V_d of each glass is

$$\frac{n_{2B} - n_{2R}}{(n_{2Y} - 1)}$$

and

$$\frac{n_{1B} - n_{1R}}{(n_{1Y} - 1)}.$$

In general, V_d or V

$$V = \frac{n_d - 1}{n_F - n_C},$$

where n_F, n_C and n_d are the refractive indices of that particular glass at the standard wavelengths 486.1, 656.3, 587.56 nm corresponding to the hydrogen F, hydrogen C and helium d lines, respectively. The V-number gives us an idea of the dispersion of a material, or in simple terms, how varied the response will be for different wavelengths. Ideally, a material should have identical refractive indices irrespective of the wavelength. In that case, the denominator of the V-number would be 0 and therefore, its value would be infinity. In reality, high values are in the 1960s or 1970s. Crown glasses have higher V-numbers and flint glasses lower values. For example, BK7 is a crown glass with V-number 64.9. Equation (4.15) can be rewritten as

$$\frac{f_{2Y}}{f_{1Y}} = -\frac{V_1}{V_2}$$

$$f_{1Y} V_1 + f_{2Y} V_2 = 0. \tag{4.16}$$

If this equation is satisfied, then our starting condition $f_R = F_B$ is valid. Since the Abbe numbers are positive (for conventional glasses) and are fixed for the glass types used, the only way this equation can be satisfied is if the focal lengths of the lenses used are of opposite sign. That is, a convex and concave lens (of different materials) need to be combined in order to reduce the chromatic aberration. Most doublets will consist of lenses made from crown and flint glasses.

4.4.4 Minimising aberrations using mirrors

In the previous section, means by which to reduce aberrations using lenses were discussed. We know that chromatic aberration occurs because of the different refractive indices seen by the different wavelengths of light propagating through the system. Therefore, systems such as spectrometers that work over a large wavelength range often use mirrors instead of lenses to redirect light. Since the light reflects off rather than transmits through the element, there is no chromatic aberration. In addition, a parabolic mirror will suffer no spherical aberration. This can be understood by keeping in mind that a parabola is defined as a path in a plane, every point of which is equidistant from a focus and a fixed line. Therefore, a collimated beam travelling parallel to the optical axis will focus on a diffraction-limited spot, as schematically indicated in figure 4.17(a). From the figure, it is clear

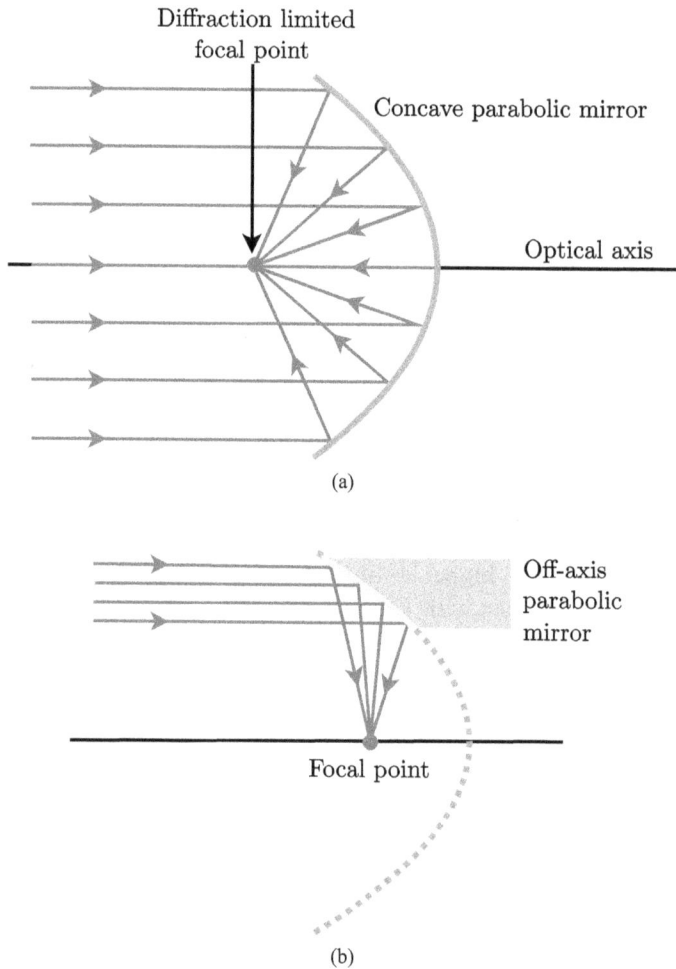

Diffraction limited
focal point

Concave parabolic mirror

Optical axis

(a)

Off-axis
parabolic
mirror

Focal point

(b)

Figure 4.17. (a) Collimated incident beam focuses to a diffracted limited spot. (b) The focal spot is separated from the incident beam and is easily accessible when using an off-axis parabolic mirror.

that the focal spot cannot be easily accessed as it is located in the path of the incident beam. Therefore, systems often use off-axis parabolic mirrors, which are created from a section of the parabola, as seen in figure 4.17(b).

In both cases, the spot is free of chromatic and spherical aberration.

4.5 Modulation transfer function

Any signal contains many different spatial frequencies in it. For example, the rays from an object can be thought of as such a signal. Aberrations help quantify the quality of an image for different rays coming from the object. The modulation transfer function (MTF), on the other hand, gives us information on how well a lens handles different spatial frequencies in the object signal. Both pieces of information are important to understand how well the lens forms an image. In general, modulation can be defined as

$$\text{modulation} = M_P = \frac{I_{\max} - I_{\min}}{I_{\max} + I_{\min}}, \tag{4.17}$$

where $I_{\max}$ and $I_{\min}$ represent the maximum and minimum intensity in the plane P. This gives an idea of how good the contrast is, in that plane. Using this definition, MTF is given by

$$\text{MTF} = \frac{M_{\text{image}}}{M_{\text{object}}}. \tag{4.18}$$

The way an imaging system affects the modulation is schematically demonstrated in figure 4.18. The object consists of a set of periodic lines (in this case, three different periods or spatial frequencies are present). The modulation of the object is 1 in all regions as $I_{\min} = 0$ and $I_{\max} = 1$.

However, after being imaged by the system, the modulation is different for each spatial frequency present. In other words, the MTF of the lens is not 1 at all frequencies. Such standard objects are available [6] and used to measure the MTF of practical systems. The spatial frequency is defined as cycles or line pairs per mm (lp mm^{-1}). A line pair consists of one bright and one dark line.

A typical MTF graph may look like the one in figure 4.19.

There are actually four curves that are typically plotted, namely the MTF for the meridional and sagittal rays at 0° and at full field. In this case, the 0° meridional and sagittal curves perfectly overlap and are, therefore, seen as a single curve.

Point to ponder: Under what circumstances, would the on-axis meridional and sagittal curves not overlap?

The full field curves do not overlap and are visible as two independent curves on the plot. The fact they do not overlap should be a clear indication that this optical

Object

Optical system

Image

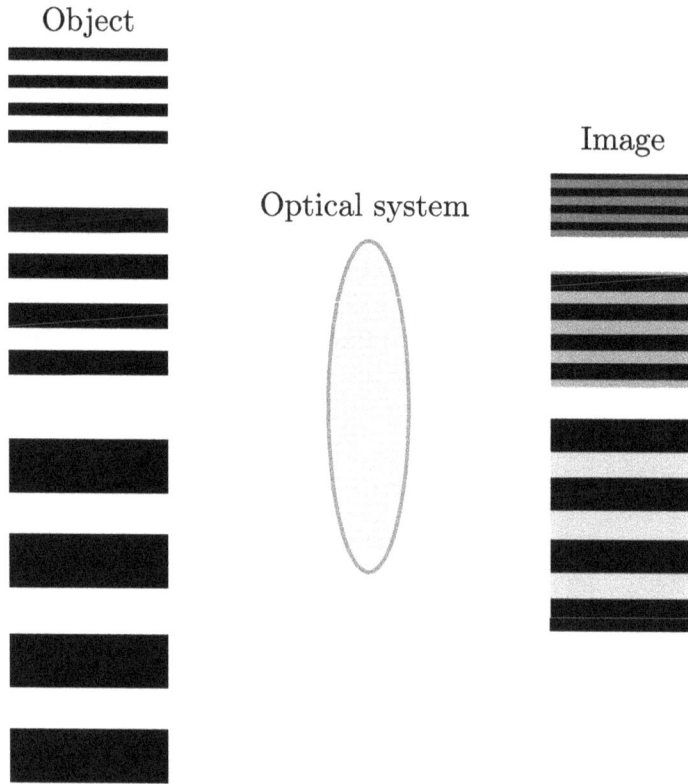

Figure 4.18. MTF.

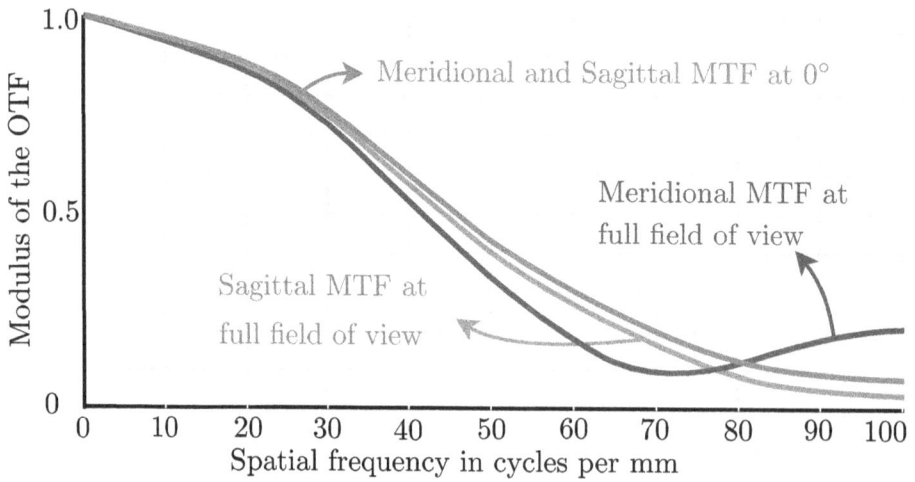

Figure 4.19. MTF.

system has astigmatism. Let us spend a moment understanding the axis labels. The y-axis is called the modulus of the optical transfer function, which is how the MTF is also referred to. It runs from 0 to 1, where 1 means the modulation of the objection is directly transferred to the image. The x-axis is given in cycles mm^{-1} or lp mm^{-1}. It should be obvious that the resolution of the optical system is closely linked to the resolution limit of the sensor being used:

$$\text{Nyquist limit} = \text{sensor resolution limit}$$

$$= \frac{1000}{2 \text{ pixel size in microns}} \text{lp } mm^{-1}. \tag{4.19}$$

For example, if a camera has a pixel of size 5 μm, its Nyquist limit will be 100 lp mm^{-1}. This is why the largest value for the x-axis of figure 4.19 is 100, as there is no point in plotting values larger than this. As a rule of thumb, if at 2/3 of the Nyquist limit, the MTF for the system is above 25%–30%, then the system will be good for imaging.

4.6 Bridging the gap between theory and design tools

In appendix A, we discuss how to design a lens of a desired focal length using either OSLO or Zemax. Optical systems can contain a number of lenses and mirrors. The main aim of the design process is to optimise the system, as discussed in section 4.4, to obtain the best results. Hence, both software offer a variety of analysis tools to study image quality, as well as the aberrations of the system.

Even casual users of these software systems will know how to optimise a lens or optical system to reduce one or more of the monochromatic aberrations. However, it is interesting to note that it is also possible to reduce chromatic aberration by 'optimising the glass type' of a lens or lenses in the system. Examples are given in section A.4 of the appendix.

4.7 Problems

1. Calculate the spherical aberration of a converging lens made of glass with a refractive index of 1.5. The lens has a focal length of 120 mm and a diameter of 50 mm. Determine the change in focal length caused by the spherical aberration.

2. Design a doublet lens that minimises chromatic aberration. The lens should have a focal length of 200 mm. Assume the convex glass has a focal length of 87.5 mm. What is the focal length of the concave lens that that will together create the doublet? For the two materials, use a Crown glass ($n_1 = 1.517$ and $V_1 = 64$) and a Flint glass ($n_2 = 1.655$ and $V_2 = 36$). Calculate some possible radii of curvature for each lens element.

3. A single lens of 20 mm diameter and EFL =65 mm has aberration curves as shown in figure 4.20. The incident beam is a monochromatic beam of 588 nm. What is the dominant aberration that is present? Curves for the same lens have been plotted in both OSLO and Zemax.

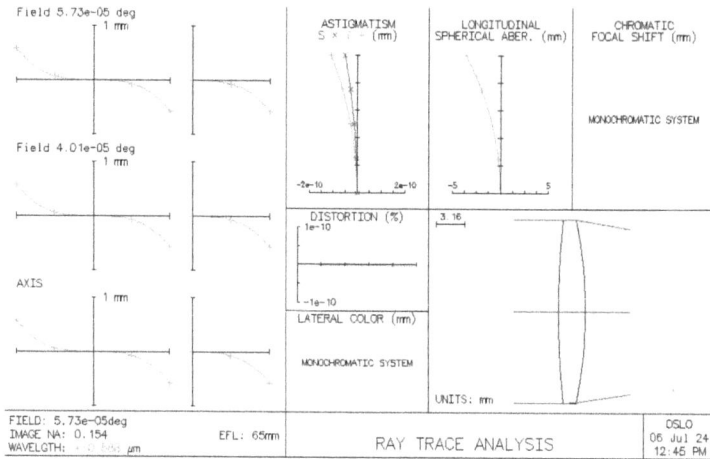

(a)

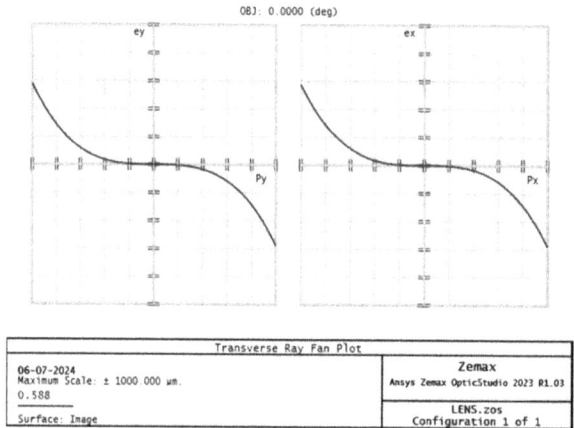

(b)

Figure 4.20. Aberrations of a single lens (a) ray intercept curves for three field values in OSLO and (b) aberration curves for identical system but plotted in Zemax. Graphs generated by SB.

4. Using any optical design tool, model a 10 mm thick bi-convex lens with radii of curvature $R_1 = 120$ mm, $R_2 = -80$ mm. Let the aperture stop be positioned 20 mm after the second surface of the lens. Make the entrance pupil diameter 10 mm and the field angle 20°. Keeping R_2 as a variable, optimise the model so that the EFL equals 100 mm. Plot the aberration curves and spot diagrams. What is the predominant aberration? What parameters can now be optimised to reduce coma without changing the focal length? Carry out this exercise and compare the aberration curves and spot diagrams before and after optimisation.

5. The MTF plots of two different optical systems, designed to work at 545 nm, each with a focal length of approximately 2.5 nm, a field angle of 25° and

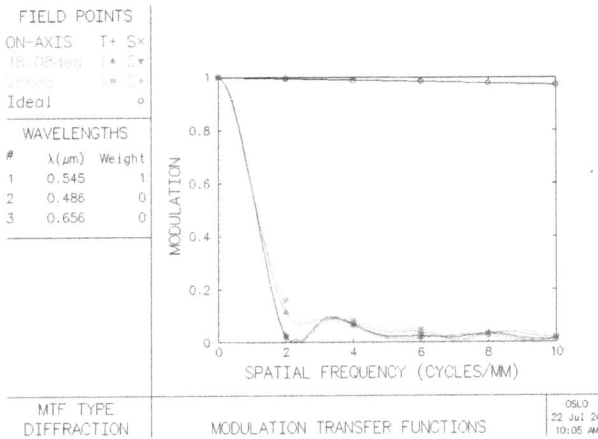

(a)

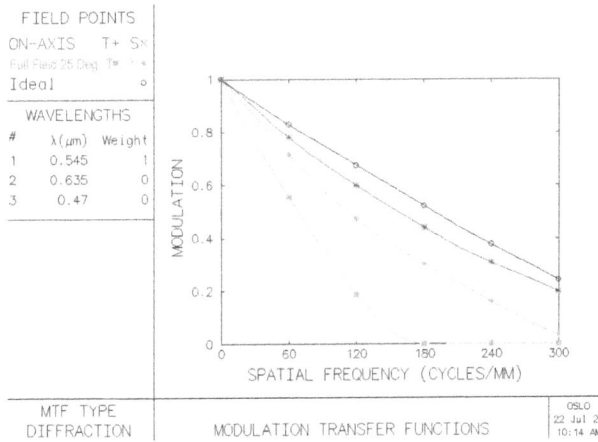

(b)

Figure 4.21. MTF curves plotted for an optical system in OSLO: (a) system 1 and (b) system 2. Graphs generated by SB.

working *f*-number of 4 are given in figures 4.21(a) and (b). One of the systems is a specially designed objective comprising a number of lenses and the other a single lens. Which graph corresponds to which system? Which would be better for imaging?

References

[1] 2015 *Aberration Theory and Correction of Optical Systems* Handbook of Optical Systems vol 3 ed H Gross (New York: Wiley)
[2] Lin P D and Johnson R B 2019 Seidel aberration coefficients: an alternative computational method *Opt. Exp.* **27** 19712–25
[3] Welford W T 1986 *Aberrations of Optical Systems* (Milton Park: Routledge)

[4] Smith W J 2000 *Modern Optical Engineering: The Design of Optical Systems* Optical and Electro-Optical Engineering Series (New York: McGraw-Hill)

[5] Oslo Optics Reference *Lambdares.com* https://lambdaresfiles.com/wp-content/uploads/support/oslo/oslo_releases/OSLOOpticsReference.pdf (Accessed: 29 March 2023)

[6] 1951 USAF resolution test chart *Wikipedia* https://en.wikipedia.org/wiki/1951_USAF_resolution_test_chart (Accessed: 6 October 2023)

Chapter 5

Gaussian beams

5.1 Gaussian beams

Until this point, we have seen how the idea of rays can be used effectively in geometric optical systems. Rays do not provide complete information about the electromagnetic field. This was not an issue when dealing with systems that used incandescent or fluorescent illumination. However, many systems and instruments today employ lasers as their light source. The most common output from such a source is a Gaussian beam, which can be considered a solution to the wave equation in a laser cavity. Kogelnik and Li [1] arrived at an expression for the Gaussian beam (GB) when looking for eigenvectors for the $ABCD$ matrices of a laser cavity. Optical cavities are typically open with reflectors on either side. If at least one mirror is converging, it is possible to produce a beam that will reproduce upon a round trip. Convergence offsets the divergence the beam acquires during propagation.

Rather than doing a formal derivation of a Gaussian beam, we look at some ideal solutions to the wave equation and use these ideas to propose a practical solution. The general wave equation is given by

$$\nabla^2 \mathbf{U}(r,\,t) - \frac{1}{c^2}\frac{\partial^2 \mathbf{U}(r,\,t)}{\partial^2 t} = 0, \tag{5.1}$$

where the wave $U(r,\,t)$ is a function of both space and time. One solution to this is

$$\mathbf{U}(r,\,t) = \mathbf{E}(r)\exp[j(-2\pi vt + \phi(r))]$$

$$= \mathbf{u}(r)\exp(-j2\pi vt). \tag{5.2}$$

Substituting (5.2) in (5.1) results in the time-independent form of the wave equation, namely the Helmholtz equation:

$$\nabla^2 \mathbf{u}(r) + k^2 \mathbf{u}(r) = 0. \tag{5.3}$$

doi:10.1088/978-0-7503-5497-4ch5

Let us pause at this point and make some observations. The solution to (5.3) should contain parts relating to the amplitude and phase of the wave. We therefore assume $\mathbf{u}(r) = \mathbf{E}(x, y, z)\exp(jkr)$. However, we started this discussion stating our interest in laser beams. Such beams are highly directional, which means that their lateral extent is very small compared to the distance they propagate over. Mathematically, we can say that $x, y \ll z$. Therefore, $r = \sqrt{x^2 + y^2 + z^2}$ in the exp term of $u(r)$ can be rewritten as $z(1 + (x^2 + y^2)/2z^2)$. This expression comprises two terms which should be familiar. The first term represents a plane wave travelling in the z direction and the other a wave with spherical behaviour.

We could push the idea of laser directionality further and write the wave equation solution as $\mathbf{u}(r) = \mathbf{E}(x, y, z)\exp(jkz)$. As the wave propagates in the z direction, the change across the wavefront is much faster than in the direction of propagation, i.e. $\partial^2\mathbf{E}/\partial^2 z \ll \partial^2\mathbf{E}/\partial^2 x, \partial^2\mathbf{E}/\partial^2 y$. In fact, we can set $\partial^2\mathbf{E}/\partial^2 z = 0$. Using these relationships in (5.3) results in what is called the paraxial wave equation:

$$\left[\frac{\partial^2}{\partial^2 x} + \frac{\partial^2}{\partial^2 y} \right]\mathbf{E} + 2jk\frac{\partial\mathbf{E}}{\partial z} = 0. \tag{5.4}$$

Until this stage, we have been surmising what the solution could be given certain observed beam properties. We now put these ideas together and propose a solution, using the idea that the solution will have a combination of plane and spherical wave behaviour. Let us say that the solution is

$$\mathbf{E}(x, y, z) = \mathbf{A}\exp\left\{ -j\left[\frac{k(x^2 + y^2)}{2q(z)} \right] \right\}\exp\{-jp(z)\}, \tag{5.5}$$

where the magnitude of A is the peak amplitude of the wave and $q(z)$ and $p(z)$ are functions governing the behaviour of the wave as it propagates in the z direction. Substituting (5.5) in (5.4) gives rise to

$$\mathbf{E}\left[\frac{k^2}{q^2(z)}\left(1 - \frac{\mathrm{d}q}{\mathrm{d}z} \right)(x^2 + y^2) + 2k\left(\frac{j}{q(z)} + \frac{\mathrm{d}p}{\mathrm{d}z} \right) \right] = 0. \tag{5.6}$$

The only way that this equation can be true for all values of x and y is if $1 - \mathrm{d}q/\mathrm{d}z = 0$ and $j/q(z) + \mathrm{d}p/\mathrm{d}z = 0$. These relationships can be be used to obtain

$$q(z) = z + jz_0 \tag{5.7}$$

and

$$p(z) = -j\ln\left(1 + \frac{z}{jz_0} \right). \tag{5.8}$$

We leave the explanation of the term z_0 for a little later in the chapter, suffice to say at this point that it represents a constant which in some way helps define a Gaussian beam. With some manipulation [2], the substitution of (5.7) and (5.8) into equation (5.5) will result in the familiar form of the Gaussian beam equation, namely

$$\mathbf{E}(x, y, z) = A\frac{w_0}{w(z)} \exp\left[-\frac{(x^2 + y^2)}{w^2(z)}\right] \exp[-jkz] \exp\left[-j\frac{k(x^2 + y^2)}{2R(z)}\right] \exp[j\phi(z)], \quad (5.9)$$

where

$$R(z) = z\left[1 + \left(\frac{z_0}{z}\right)^2\right], \quad (5.10)$$

$$w(z) = w_0\sqrt{1 + \left(\frac{z}{z_0}\right)^2}, \quad (5.11)$$

$$w_0 = \sqrt{\frac{z_0\lambda}{\pi}}, \quad (5.12)$$

and

$$\phi(z) = \tan^{-1}\left(\frac{z}{z_0}\right). \quad (5.13)$$

The significance of these equations and terms (including the definition of w_0) will be explored in subsequent sections. However, it should be obvious that the part in blue in equation (5.9) relates to the amplitude of the beam at a particular z plane. The equations are for a beam of wavelength λ.

5.2 Gaussian beam properties

From equations (5.10) to (5.13), it should be obvious that the different parameters [3] describing a GB are quite interlinked. It is difficult to choose which one to start explaining, as they each require an understanding of all the other terms! We take the leap by looking at the intensity of Gaussian beams first.

5.2.1 Optical intensity

The Gaussian beam derives its name from the Gaussian spread of the intensity distribution. As the beam propagates, the form of the intensity cross section at any plane remains Gaussian. The intensity can be obtained from equation (5.9) using

$$I(x, y, z) = |\mathbf{E}(x, y, z)|\, |\mathbf{E}^*(x, y, z)|$$

$$= |\mathbf{A}|^2\left(\frac{w_0}{w(z)}\right)^2 \exp\left\{-\left[\frac{2(x^2 + y^2)}{w^2(z)}\right]\right\}$$

or

$$I(\rho, z) = |\mathbf{A}|^2\left(\frac{w_0}{w(z)}\right)^2 \exp\left\{-\left[\frac{2\rho^2}{w^2(z)}\right]\right\}, \quad (5.14)$$

where $x^2 + y^2$ has been replaced by ρ^2.

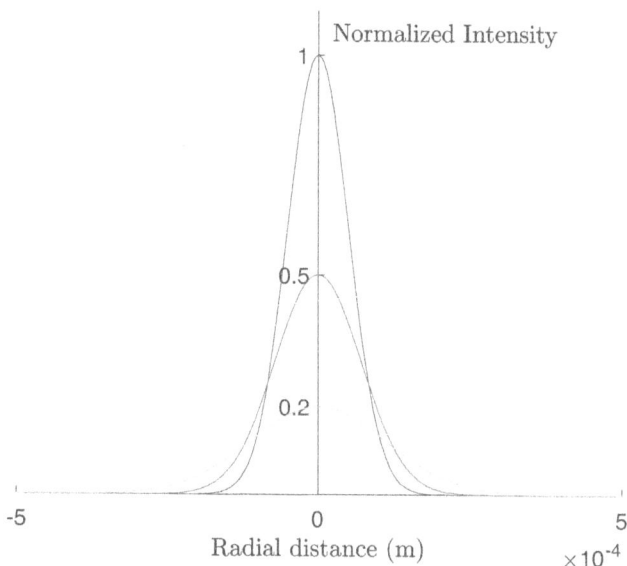

Figure 5.1. Intensity across a GB at different z planes.

Figure 5.1 shows the intensity cross section of a beam at three different locations, $z = 0$, $z = z_0$ and $z = 2z_0$. While the shape remains the same, the central peak drops drastically.

5.2.2 Beam radius

Scientists need to be able to measure quantities. Doing so enables an accurate comparison of different beams or helps in more efficient and better designs of optical systems that will use those beams. Whatever the reason, clearly, we must be able to assign a size to a GB. If the beam had a finite boundary, we could define its diameter in terms of that. But the edges of a Gaussian beam extend all the way to infinity. Its size is therefore expressed relative to its peak intensity. The radius $w(z)$ at any position z is defined as the width at which the intensity has fallen to $1/e$ of its peak intensity and is given by equation (5.11). w_0 is the beam radius at the narrowest point of the beam and is called the beam waist. In the case of a beam emerging from a laser, the waist would be the radius at the exit of the laser, as the beam would immediately start diverging after leaving the laser cavity. Figure 5.2 shows how the beam radius of two beams varies as the beam propagates. Both have the same waist size at $z = 0$ but because of their different wavelengths, diverge quite differently. Since the radius is initially large, then takes on a minimum value and then expands again, it is clear that the GB has been focused by an optical element.

5.2.3 Power

It can often be useful to express the intensity of a beam in terms of its power, especially as power is the quantity that scientists typically measure. As long as the

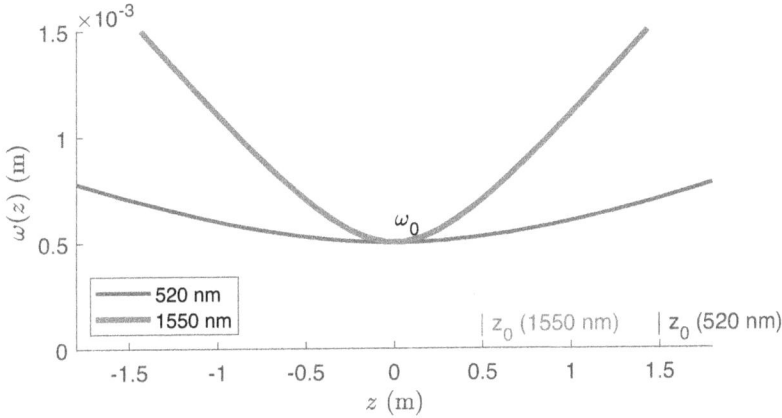

Figure 5.2. Difference in beam divergence for two Gaussian beams with same initial waist size but generated by different wavelength sources.

medium of propagation is considered non-absorbing and lossless, the power should be the same in any plane. It can be calculated by integrating equation (5.14) from 0 to ∞,

$$P = \int_0^\infty I(\rho, z)2\pi\rho \mathrm{d}\rho$$

which results in

$$= \frac{1}{2}I_0(\pi w_0^2). \tag{5.15}$$

The expression within brackets can be considered to be an *area* that represents the GB. Equation (5.14) can then be rewritten in terms of the power as

$$I(\rho, z) = \frac{2P}{\pi w^2(z)} \exp\left[\frac{-2\rho^2}{w^2(z)}\right].$$

The ratio of power within a beam of radius ρ_0 to the total power is given by

$$\frac{1}{P}\int_0^\infty I(\rho, z)2\pi\rho \mathrm{d}\rho = 1 - \exp\left[\frac{-2\rho_0^2}{w^2(z)}\right]. \tag{5.16}$$

We can now use this expression to judge how good our definition of GB size really is. Ideally, 100% of the power should lie within a circle defined by the GB radius. By setting $\rho_0 = w(z)$ in (5.16), we find that only 86% of the power lies within this circle. This gives rise to an important point to be considered when designing Gaussian beam optics. In order to efficiently capture and utilise a Gaussian beam, the size of the optics must be larger than that specified by the beam diameter. This is true even if the light is going to fall normally incident on the optics. Typically, a *safety factor of at least 2* is used when choosing diameters of optics with respect to GB beam sizes.

5.2.4 Depth of focus

The idea of the DOF was presented in chapter 2. In an imaging system, it was defined as the distance over which the image plane could be shifted with a minimal or acceptable degradation in image quality. The depth of focus was not discussed when considering a plane wave travelling, since in ray optics, an ideal collimated beam of light would retain its size even if it propagated to infinity. One could think of its depth of focus as being infinite. However, Gaussian beams represent practical beams that always undergo diffraction, resulting in divergence. There are no perfectly collimated GBs! The DOF of a GB can be related to the rate of divergence or how quickly the beam deviates from its beam waist. A measure of this is given by the Rayleigh range z_0 of a beam which can be obtained from equation (5.12)

$$z_0 = \frac{\pi w_0^2}{\lambda}. \tag{5.17}$$

The DOF of a GB can be considered to be the distance over which the change in the beam radius is small enough for the beam to be considered still collimated. This distance is conventionally defined as $2z_0$ and the radius at the end of the DOF is obtained by substituting at $z = z_0$ in equation (5.11), which results in $w(z_0) = \sqrt{2}\,w_0$.

5.2.5 Beam divergence

As discussed, over the distance $-z_0$ to z_0, the waist is slowly varying. Beyond z_0, the rate of change of beam radius with respect to z increases and for larger z values it can almost be considered a linear or asymptotic variation. By convention, this variation is called the beam divergence and is defined as

$$\theta_0 = \frac{w_0}{z_0}. \tag{5.18}$$

Substituting equation (5.17) in this equation results in a half-angle definition of divergence that can be given by

$$\theta_0 = \frac{2\lambda}{(2w_0)\pi}. \tag{5.19}$$

The 2 is explicitly mentioned in the denominator to make it clear that increasing the beam diameter $(2w_0)$ would decrease the beam divergence and vice versa. Two beams with the same waist would diverge differently if they were generated by different sources.

Point to ponder: To have a Gaussian beam travel a long distance, is it better to start off with a beam with a small or fat waist?

5.2.6 Beam quality

We started discussing Gaussian beams as they are practical beams that emerge from lasers as opposed to the plane waves that are normally used to introduce the idea of waves. However, practical lasers will not all generate perfect Gaussian beams. The M^2 parameter is used to measure how closely a real Gaussian beam is to an ideal one [4]. It is defined as

$$M^2 = \frac{\theta_0'}{\theta_0} = \frac{\text{actual divergence}}{\text{divergence of ideal GB with same waist}}$$

$$= \frac{\pi \theta_0' 2\omega_0}{2\lambda}. \tag{5.20}$$

One can think of the M^2 parameter as an inherent property that the beam is imbued with when it is generated. It cannot be improved. If optics is used to alter the beam waist, then the beam divergence will change in such a way that the product

$$\omega_0 \theta_0 \tag{5.21}$$

stays constant. This product is called the beam parameter product (BPP) and its magnitude is of the order of $O(\frac{\lambda}{\pi})$. Both the BPP and M^2 parameter provide information on how well a beam will focus. An ideal GB will have $M^2 = 1$ and a non-GB will be M^2 larger than a perfect beam in the far field. This relation between beam waist and divergence provides a result similar to the uncertainty principle. The better localised the photons are in space (that is a beam with a small waist), the larger the uncertainty in their angle of propagation (i.e. the greater their divergence).

5.2.7 Gouy phase

Earlier on in the book, we discussed how the phase of light changes as it propagates. There are other methods for the phase to change as well. One such method is through something called the Gouy phase [5, 6]. The Gouy phase shift is a phase change that a converging beam experiences when it passes through its focus, or in the case of a Gaussian beam, through its beam waist. If a Gaussian beam was compared to a plane wave of the same wavelength and propagating in the same direction, the Gouy phase would mean that there would be a phase difference between the GB and the plane wave. The total phase change as the beam travels from $-\infty$ to $+\infty$ is π radians, as can be seen in figure 5.3.

Point to ponder: How can the Gouy phase be understood in terms of the velocity of the wave?

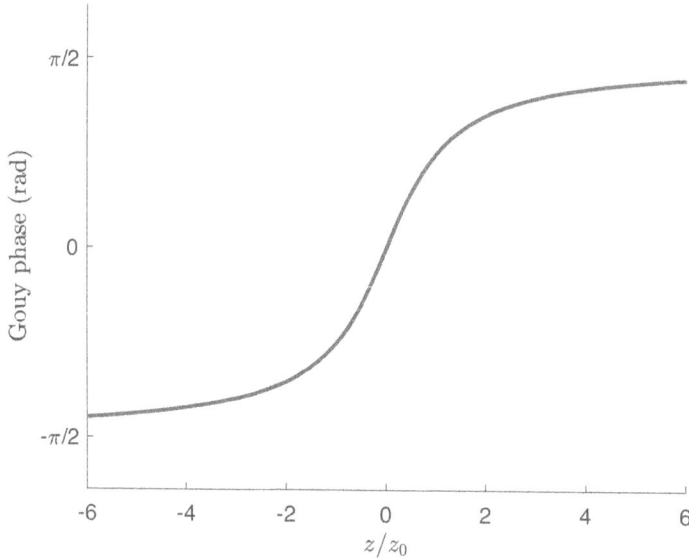

Figure 5.3. The Gouy phase.

5.3 Characterising a Gaussian beam

Each solution to the Helmholtz wave equation can be characterised or completely defined by a specific set of parameters. The plane wave can be characterised by its amplitude and the direction of propagation, the spherical wave by its amplitude and origin. The GB requires a few more parameters, namely its peak amplitude, direction of travel, location of waist and either the value of ω_0 or z_0. It turns out that several of these terms can be addressed through a single parameter called the complex parameter q of the GB. It is defined as

$$\frac{1}{q(z)} = \frac{1}{R(z)} - j\frac{\lambda}{\pi\omega^2(z)}.$$ (5.22)

This can be inverted to obtain q

$$q = \frac{1}{\left(\dfrac{1}{R(z)} - j\dfrac{\lambda}{\pi\omega^2(z)}\right)}.$$

Replacing $R(z)$ and $w(z)$ with equations (5.10)–(5.12) gives rise to

$$q = \frac{1}{\left(\dfrac{z}{z^2 + z_0^2} - j\dfrac{z_0}{z^2 + z_0^2}\right)},$$

which on simplification becomes

$$= z + jz_0.$$ (5.23)

A close look at equation (5.22) indicates that:
1. The real part of $q(z)$ provides the waist location.
2. The imaginary part is the Rayleigh range.
3. From z_0, ω_0 can be obtained.
4. $w(z)$ can be calculated from ω_0.
5. Interestingly, the argument of q is $\tan^{-1}(z_0/z)$, which is nothing other than the Gouy phase of the beam.

In other words, all information about the GB is stored in q.

5.4 Transmittance of an optical element

We already know that these parameters change as the beam travels through an optical system. The question to ask at this point is then, given the values of the GB at one location of an optical system, how can we find out its values at another location. Figure 5.4 demonstrates this change for the simple case of a beam travelling through a lens.

We will assume that the lens is perfectly transmissive so that the amplitude of the beam does not change as it travels through the lens. In that case, the phase of the beam will be altered in accordance with the phase of the lens.

5.4.1 Transmittance of an element

The transmittance of an element is the ratio of the field immediately after the element $u_d(x, y)$ to the incident field $u_0(x, y)$. It is given by

$$t(x, y) = \frac{u_d(x, y)}{u_0(x, y)}. \tag{5.24}$$

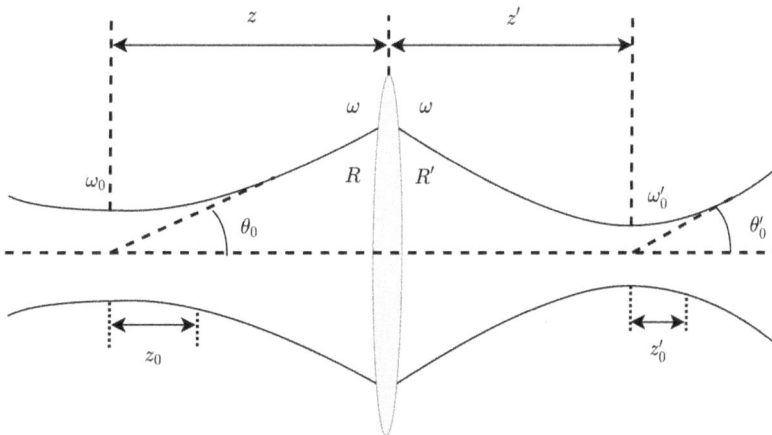

Figure 5.4. Transformation of a Gaussian beam as it travels through a lens.

If the incident wave is assumed to be a plane wave of wavelength λ travelling through a slab of glass of thickness d and refractive index n, the phase change experienced across the beam is $2\pi/\lambda nd$ or $k_0 nd$. The transmittance of the element is

$$t(x, y) = \exp(-jk_0 nd). \tag{5.25}$$

What happens if the glass slab is no longer uniformly thick but has a thickness $d(x, y)$, as shown in figure 5.5(a)?

It should be clear that the beam at any (x, y) location travels through a distance $d_0 - d(x, y)$ in air and $d(x, y)$ in the slab. Therefore, the total phase acquired is

$$t(x, y) = \exp[-jk_0 nd(x, y)]\exp[-jk_0(d_0 - d)]$$

$$= \exp[-jk_0 d_0]\exp[-jk_0(n - 1)d(x, y)], \tag{5.26}$$

where the first exponential is a constant factor. We can next look at a glass slab with a specific shape, e.g. a plano-convex shape as given in figure 5.5(b).

The thickness of the lens is

$$d(x, y) = d_0 - PQ$$

$$= d_0 - (R - \sqrt{R^2 - (x^2 + y^2)}).$$

When $x, y < R$, the expression reduces to

$$= d_0 - \frac{x^2 + y^2}{2R}. \tag{5.27}$$

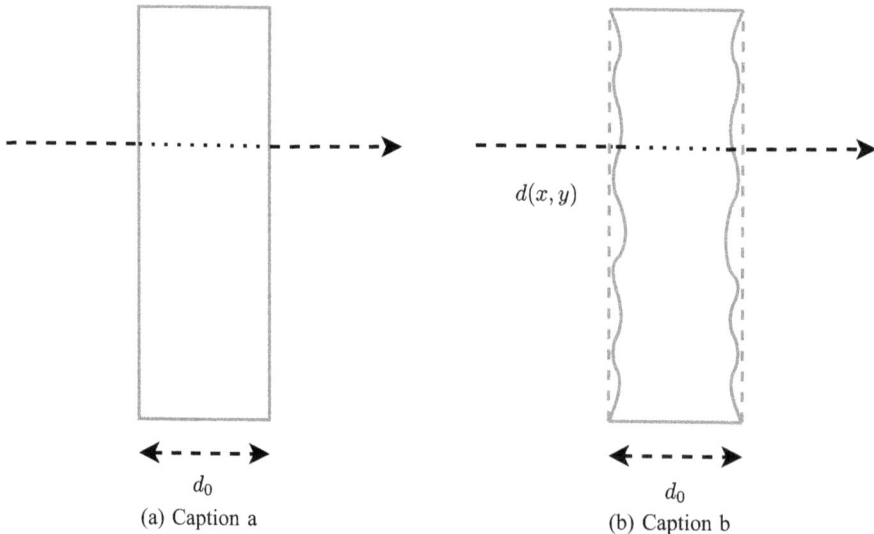

$d(x,y)$

d_0

d_0

(a) Caption a

(b) Caption b

Figure 5.5. (a) Uniform glass slab and (b) glass slab with varying thickness.

Substituting this into equation (5.26) gives the transmittance of this lens as

$$t(x, y) = \exp[-jk_0 d_0] \exp\left[-jk_0(n - 1)\left(d_0 - \frac{x^2 + y^2}{2R}\right)\right].$$

By clubbing all the constants into one term h_0, the equation can be reduced to

$$t(x, y) = h_0 \exp\left[-jk_0 \frac{x^2 + y^2}{2f}\right], \tag{5.28}$$

as $1/f = (n - 1)/R$.

Using these ideas, it should be clear that a GB phase will be altered by the phase of the lens in the following way:

$$\left(kz + \frac{k\rho^2}{2R(z)} - \zeta\right) - \frac{k\rho^2}{2f} = \left(kz + \frac{k\rho^2}{2R'(z)} - \zeta\right),$$

which can be simplified to

$$\frac{1}{R'} = \frac{1}{R} - \frac{1}{f}. \tag{5.29}$$

5.5 Matrix methods for Gaussian beams

We have seen how travelling through a transparent optical element changes the phase of a beam. In the process, the behaviour of the beam is also modified. Since we know that the complex parameter is a complete representation of a GB, any transformation of the beam must result in a change in q as well.

5.5.1 Change of q with propagation

Consider a GB with

$$q_1 = z_1 + jz_0.$$

After propagating some distance, its q parameter would be

$$q_2 = z_2 + jz_0$$

$$= q_1 + \Delta z, \tag{5.30}$$

where $\Delta z = z_2 - z_1$ is the distance travelled by the beam.

5.5.2 Change of q with transmission through a lens

From equation (5.29), it is clear that the radius of curvature of a GB is transformed by the focal length of the lens it is travelling through. Therefore, given that the q parameter follows this equation, (5.22), it should be altered such that

$$\frac{1}{q'} = \frac{1}{q} - \frac{1}{f}.$$

This can be rewritten as

$$q' = \frac{q}{\frac{-q}{f} + 1}. \tag{5.31}$$

5.5.3 Gaussian beams and the *ABCD* matrix

At this point, let us take a leap of faith and assume that propagation matrices that were used in geometric optics are valid for GBs as well. Let us further assume that an *ABCD* matrix would change the q parameter as follows:

$$q' = \frac{Aq + B}{Cq + D}. \tag{5.32}$$

In the case of a GB travelling a distance Δz, the new q would be given by

$$\begin{pmatrix} q_2 \\ 1 \end{pmatrix} = \begin{pmatrix} 1 & \delta z \\ 0 & 1 \end{pmatrix} \begin{pmatrix} q_1 \\ 1 \end{pmatrix}, \tag{5.33}$$

which is nothing other than (5.30). Similarly, transformation through a lens of focal length f would result in (5.31).

It might seem strange that the q parameter can be transformed using the *ABCD* matrices

$$z - z_c = \frac{y}{\tan \theta} \approx \frac{y}{\theta}, \tag{5.34}$$

but

$$\begin{pmatrix} y_2 \\ \theta_2 \end{pmatrix} = \begin{pmatrix} A & B \\ C & D \end{pmatrix} \begin{pmatrix} y_1 \\ \theta_1 \end{pmatrix}. \tag{5.35}$$

From the two equations, we see that

$$z - z_c \approx \frac{\text{top element of equation (5.35)}}{\text{bottom element}}. \tag{5.36}$$

Given that q can be written in terms of z, this use of the *ABCD* matrices should be a little more clear.

5.5.4 Case study: propagation through a lens

We will study this case by looking at different Gaussian beam parameters at three different planes. Plane 1, which can be considered to be an input plane, plane 2, where the thin lens lies, and plane 3, the output plane. Assuming that the waist of the GB is incident on plane 1, $\omega = \omega_{01}$ and $R_1 = \infty$. This means that equation (5.22) reduces to

$$\frac{1}{q_1} = \frac{-j}{z_{01}}.$$ (5.37)

We know that after the lens, the q parameter will change according to (5.31):

$$q_2 = \frac{\pi f \omega_{01}^2 \left(-\pi \omega_{01}^2 + j\lambda f\right)}{(\pi \omega_{01}^2)^2 + (\lambda f)^2}.$$ (5.38)

If this is written in the form $a + jb$, then

$$a = \frac{-f(\pi \omega_{01}^2)^2}{(\pi \omega_{01}^2)^2 + (\lambda f)^2}$$ (5.39)

and

$$b = \frac{f^2 \lambda \pi \omega_{01}^2}{(\pi \omega_{01}^2)^2 + (\lambda f)^2}.$$ (5.40)

The remaining operation is the distance d that the light has to travel to reach plane 3:

$$\begin{aligned} q_3 &= q_2 + d \\ &= (a + d) + jb \end{aligned}$$ (5.41)

or

$$\frac{1}{q_3} = \frac{1}{R_3} - j\frac{\lambda}{\pi \omega_3^2}.$$ (5.42)

From equation (5.41) we obtain

$$R_3 = \frac{(a + d)^2 + b^2}{a + d}$$ (5.43)

and

$$\frac{\lambda}{\pi \omega_3^2} = \frac{b}{(a + d)^2 + b^2}.$$ (5.44)

In the special case that the beam is focused at this distance, i.e. a beam waist is formed at plane 3, then $a + d = 0$, which results in

$$\frac{\omega_{03}^2}{\omega_{01}^2} = \frac{1}{1 + (z_0/f)^2},$$ (5.45)

that shows the final beam waist is controlled by the initial parameters and the lens focal length as expected.

The interesting result is obtained when solving for d from $a + d = 0$ directly:

$$d = \frac{f}{1 + (f/z_0)^2} < f.$$ (5.46)

This example demonstrates a vital difference between a GB and a plane wave. In geometric optics, the latter would always come to focus at a distance corresponding to the focal length but this example shows that a GB comes to focus closer to the lens. Even though the incident beam had a *plane wave nature*, the GB is not a plane wave.

However, if we place the additional constraint that $2z_0 \gg f$, then (5.46) reduces to the more expected result that $d \approx f$. In other words, imaging occurs at the focal plane. Why does this occur in this case and not in the earlier one? To answer that, we need to understand what the condition $2z_0 \gg f$ means. Clearly, the depth of focus of this beam is much greater than the focal length. The lens *sees* the incident beam as a plane wave, better mimicking geometric optic conditions, resulting in a familiar result.

This additional constraint changes equation (5.45)

$$\omega_{03} = \frac{\lambda f}{\pi \omega_{01}} = \theta_{03} f. \tag{5.47}$$

Rewriting this in terms of beam diameters of the incident and final beams, we obtain

$$2\omega_{03} = \frac{4\lambda f}{\pi(2\omega_{01})} = \theta_{03} f$$

$$= \frac{4}{\pi} \lambda F_{\#}$$

$$= 1.27 \lambda F_{\#}. \tag{5.48}$$

Equation (5.48) once again highlights a difference between GBs and geometric optics, as the focused GB spot is seen to be different from the one obtained in equation (2.16), which originates solely from the circular nature of the optics involved.

5.6 Gaussian beam transformation through a lens

A lens transforms a GB, as shown in figure 5.4, causing a change in the beam parameters [3]. These changes can be written in terms of the magnification M.

The new waist radius is

$$\omega_0' = M\omega_0. \tag{5.49}$$

The waist location changes according to

$$(z' - f) = M^2(z - f). \tag{5.50}$$

The depth of focus or Rayleigh range

$$2z_0' = M^2 2z_0. \tag{5.51}$$

Divergence

$$2\theta_0' = \frac{2\theta_0}{M}. \tag{5.52}$$

Magnification

$$M = \frac{M_r}{\sqrt{1+r^2}}, \tag{5.53}$$

where $r = z_0/(z-f)$ and $M_r = |f/(z-f)|$.

In section 5.5.4, we saw that even though the equivalent of a plane wave was incident on a lens, focusing occurred at a different place compared to what would happen in geometric optics. We might ask if there is no situation in which a GB has a more ray-optics behaviour.

Let us explore what happens when the lens is well outside the Rayleigh range, or in other words, $z - f \gg z_0$. In that case, from (5.53), we see that $r \ll 1$ and $M \approx M_r$. Substituting these into (5.50) results in

$$z' - f = \left(\frac{f}{z-f}\right)^2 (z-f),$$

which can be rewritten as

$$\frac{1}{f} = \frac{1}{z'} + \frac{1}{z}. \tag{5.54}$$

This equation is similar to equation (2.10), which we explored in the second chapter.

In this chapter, we stepped away from ray optics and started considering light as a wave and yet found conditions under which geometric optical behaviour was still present. This chapter is the basis of understanding the wider topic of complex light, which is discussed in chapter 8.

5.7 Problems

1. A 10 mW green laser of wavelength $\lambda = 540$ nm produces a Gaussian beam with a spot size $2w_0 = 0.04$ mm. (a) Determine the Rayleigh range of the beam. (b) What is the radius of curvature of the beam at $z = 0$, and at distances equal to z_0, $2z_0$?
2. A 20 mW 1550 nm laser produces a Gaussian beam with a spot size of $2w_0 = 10$ μm. How large a detector would one need on a geostationary satellite, if one wanted to collect 20% of the power? Assume the distance of satellite to be 35 786 km.
3. What radius should a circular aperture have such that 75% power is transmitted for a Gaussian beam with $2w_0 = 10$ μm?
4. A Gaussian beam is incident on a lens of focal length 5 cm. The distance of the waist to the lens is 15 cm. The beam travels until it reaches a concave mirror 5 cm away, tilted at 45° to the optical axis. The mirror has a radius of curvature of 1 m. Using the concept of *ABCD* matrices, calculate the following:
   (a) $q(z)$ at the beam waist in object space.
   (b) The *ABCD* matrix of this system.

(c) $q(z)$ at a screen 15 cm away from the mirror. Use the above matrix to calculate this.

(d) If the light has wavelength 633 nm, what is the spot size ω_0 of the beam in the region between the mirror and screen.

(e) Verify the values of the real and imaginary parts of $q(z)$ at the screen using only relevant equations and not the *ABCD* matrix. Assume the beam waist in object space to be 1 mm.

5. At some plane, a Gaussian beam of wavelength 550 nm has a q parameter given by $-0.0038 + j0.0132$. At what distance from this plane does the beam waist occur and what is its size?

References

[1] Kogelnik H and Li T 1966 Laser beams and resonators *Appl. Opt.* **5** 1550–67
[2] Sirohi R S 2016 *Introduction to Optical Metrology* (Boca Raton, FL: CRC Press)
[3] Saleh B E A and Teich M C 2019 *Fundamentals of Photonics* (New York: Wiley)
[4] Siegman A E 1998 How to (maybe) measure laser beam quality *DPSS (Diode Pumped Solid State) Lasers: Applications and Issues* (Washington, DC: Optica) p MQ1
[5] Feng S and Winful H G 2001 Physical origin of the Gouy phase shift *Opt. Lett.* **26** 485–7
[6] Lee T, Cheong Y, Baac H W and Guo L J 2020 Origin of Gouy phase shift identified by laser-generated focused ultrasound *ACS Phot.* **7** 3236–45

Introduction to Ray, Wave, and Beam Optics with Applications

Shanti Bhattacharya

Chapter 6

Basics of interference

We are now at the point of the book where we once again remind ourselves that light is a wave and we explore the effects of the phase of the wave in more detail. Our focus in this chapter is studying the effect called interference. While interference is harder to explain than geometric optics, its effects are seen all the time in everyday life. Some examples are the colours of soap bubbles, an oil patch on the ground or of spectacles with anti-reflection coatings on them.

6.1 Theory

To introduce the idea of interference, we will start with the very simple case of two beams of light interfering. The electric field of each beam can be written as

$$\vec{E}_1(\vec{r},\, t) = \vec{A}_1 \cos(wt - \vec{k}_1 \cdot \vec{r} + \alpha_1) \tag{6.1}$$

and

$$\vec{E}_2(\vec{r},\, t) = \vec{A}_2 \cos(wt - \vec{k}_2 \cdot \vec{r} + \alpha_2), \tag{6.2}$$

where $|A_i|$ is the amplitude of the ith beam, polarised according to the vector $\vec{A}_i$. $\vec{k}_i$ is the wave vector and α_i is an arbitrary phase. The angular frequency of both beams is w. Each term of the cosine is in radians. We are interested in the vector addition of these two fields, i.e. $\vec{E}(\vec{r},\, t) = \vec{E}_1 + \vec{E}_2$. While it is technically correct to discuss interference in terms of the fields themselves, one should note that no practical detector (and that includes the human eye) can respond to the speed that light oscillates at (10^{14} in the visible regime). All currently existing detectors give us a time average value of the field rather than the field itself. This term, called the intensity of the beam, can be arrived at by taking the time average of the Poynting vector $\vec{S}$ of the field with amplitude A and is given by

$$I = \langle \vec{S} \rangle = \frac{c\epsilon_0 n}{2} A^2, \tag{6.3}$$

doi:10.1088/978-0-7503-5497-4ch6

where c is the velocity of light in air, ϵ_0 is the permittivity of free space and n is the refractive index of the medium the light is travelling in. For the most part, we will assume it is 1. The units of intensity are W/m^2 or the optical power per unit area. It can therefore, be understood to represent the power that travels through a surface perpendicular to the direction of propagation. Also, the constants are usually neglected in discussions regarding intensity, especially when we are more concerned about the relative values across a plane. Keeping this in mind, what we are interested in then is the term $E^2 = \vec{E} \cdot \vec{E}$ given by

$$E^2 = E_1^2 + E_2^2 + 2\vec{E}_1 \cdot \vec{E}_2. \tag{6.4}$$

This can be written in terms of the intensities

$$I = I_1 + I_2 + 2\sqrt{I_1 I_2} \cos \delta, \tag{6.5}$$

where $\delta = \alpha_1 - \alpha_2 + \vec{k}_1 \cdot \vec{r} - \vec{k}_2 \cdot \vec{r}$, which represents the phase difference between the two interfering beams. A more useful form of the interference equation is

$$I = I_0(1 + V \cos \delta), \tag{6.6}$$

where the visibility or modulation depth is given by $V = 2\sqrt{I_1 I_2}/(I_1 + I_2)$ and the dc or average value of the intensity by $I_0 = I_1 + I_2$. Equation (6.6) tells us that the interference of two beams would result in an interference pattern with fringes, as shown in figure 6.1.

When the phase difference between the beams is 0 or an even multiple of π, i.e. $\delta = 2m\pi$, constructive interference takes place and equation (6.6) reduces to $I = I_0(1 + V)$. On the other hand, when $\delta = (2m + 1)\pi$, destructive interference occurs and equation (6.6) reduces to $I = I_0(1 - V)$. No interference takes place if $\delta = (2m + 1)\pi/2$. The clearest fringes (i.e. fringes with the best contrast or visibility) are seen when $I_1 = I_2$ or in other words $V = 1$. This should not be surprising. If the intensities of the interfering beams were very different, such that $I_1 \ll I_2$, it would be as if only one beam was playing a role and therefore, one would not expect interference to take place and any fringes formed would have poor visibility. Figure 6.2 shows the variation in the modulation or visibility of fringes formed for the cases of $V = 0$, 0.5 and 1, respectively.

Figure 6.1. Fringes when two plane waves interfere.

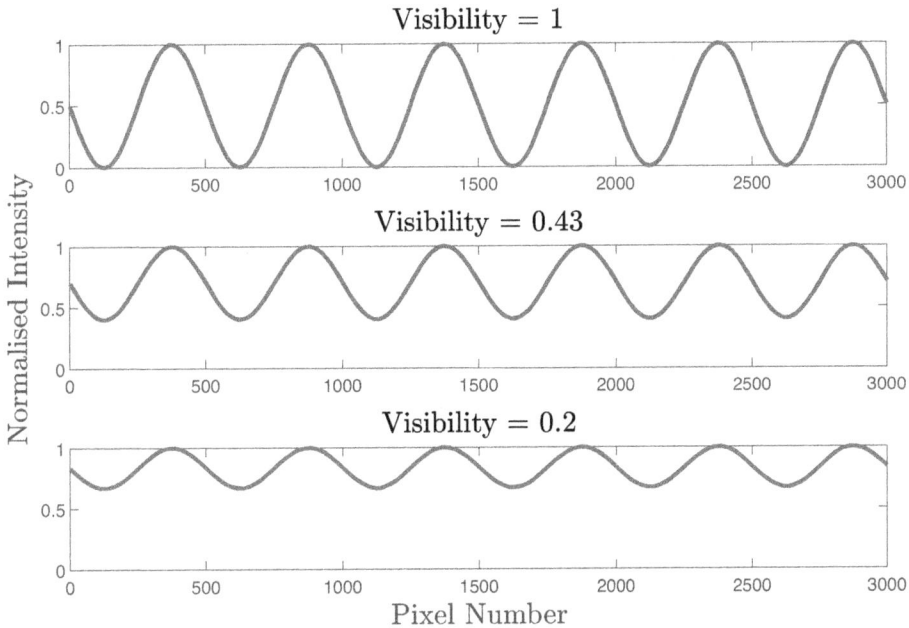

Figure 6.2. Variation in modulation versus intensity of of interfering beams.

6.2 Conditions for interference

From the previous discussion, it might appear that mixing any two light beams together always results in interference. This is far from the truth, however. Interference occurs under some very specific conditions [1]. We see that the term of importance in equation (6.6) contains the phase difference between the two interfering beams at that point of time. If the phase difference was randomly changing, interference would not take place. In essence, there has to be a specific or fixed phase relationship between the interfering beams for interference to take place. Given the way light is generated by a source, it is impossible for the phase of beams from two different sources to be correlated. But even when the beams come from one source, interference is not guaranteed. Interference can be observed relatively easily when using a coherent source. A light field can be considered coherent when there exists a fixed phase relationship between the electric field values at different spatial positions or at different times. Light sources can also be partially coherent if there exists some correlation between phase values. To understand coherence better we look at it from two different aspects, keeping in mind that we are talking about beams or electric fields generated from a single source.

6.2.1 Spatial coherence

Spatial coherence relates to the phase relationship between the electric field at two different points but at the same instant of time. The famous Young's double slit experiment [2] can be done to ascertain the spatial coherence of a source. The set-up is shown in figure 6.3.

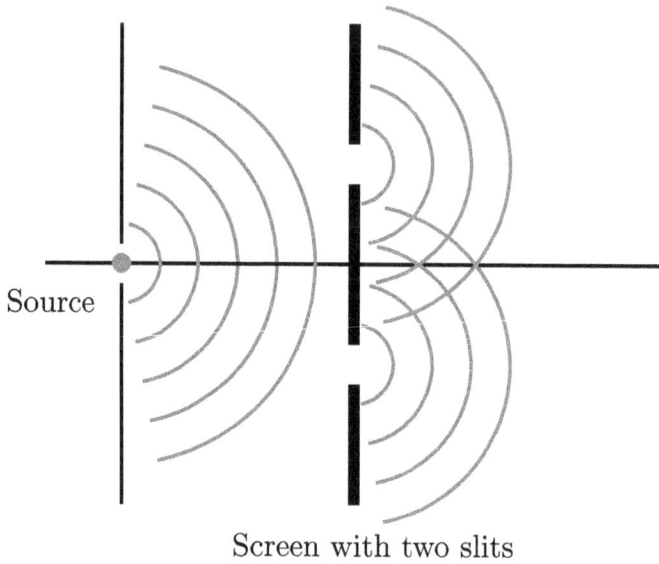

Screen with two slits

Figure 6.3. Young's double slit experiment.

Since coherence relates to phase, and phase is not directly measurable, this experiment allows us to see the impact of spatial coherence through the fringes formed. A point source would have perfect spatial coherence. As the size of the source increases, the spatial coherence decreases.

6.2.2 Temporal coherence

A Michelson interferometer [3], a schematic of which is shown in figure 6.4, can be employed to study the temporal coherence of a light source. A beam has good temporal coherence if the phase relationship of an electric field at two different points of time, at one spatial location, is fixed.

The temporal coherence of a source is related to its linewidth. To appreciate why this is, let us take the case of a HeNe laser operating at a wavelength $\lambda = 632.8$ nm, with a bandwidth $\Delta\lambda = 2 \times 10^{-3}$ nm. Given the relationship where the velocity of light in free space is given by $c = \nu\lambda$, we can find the (frequency) bandwidth of the source from

$$\Delta\nu = \frac{c\Delta\lambda}{\lambda^2}. \tag{6.7}$$

Any phase relationship is only constant in a time period less than Δt_c, where $\Delta t_c = 1/\Delta\nu$, i.e. the inverse of the term given in equation (6.7). For the case of the HeNe laser, $\Delta\nu = 1.5$ GHz. The implication of this is that every $\Delta t_c = 0.666 \times 10^{-9}$ s, the wavelength shifts to some value within $\lambda \pm \Delta\lambda/2$.

To make Δt_c larger, $\Delta\nu$ should decrease. For the HeNe laser, $\Delta\nu$ was in the order of GHz but it would have to be only 1 Hz to have the phase maintained for just 1 s! This should also help explain why interference cannot be observed between beams from independent sources, as it is impossible to maintain their phase difference. Even for lasers, interference will only be observed over a distance travelled by the beam in the time Δt_c. This distance is called the coherence length of source and is given by

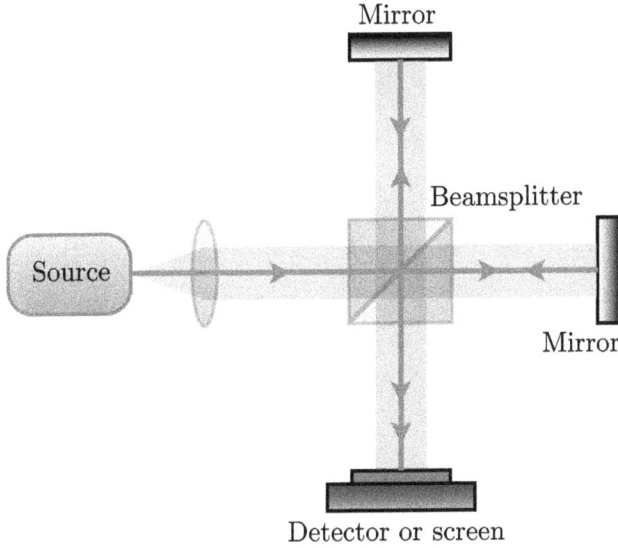

Figure 6.4. Michelson interferometer.

$$L_C = c\Delta t_c. \tag{6.8}$$

Lasers generate light over a very small region and from a very narrow bandwidth, which is why they generally have extremely high spatial and temporal coherence.

6.2.3 Polarisation

Equation (6.5) was arrived at assuming that the polarisation of the interfering beams was identical. This means that in each case, the full field (given by equations (6.1) and (6.2)) plays a role in the interference. The dot product of equation (6.4) then depends only on the phase difference between the two beams.

If the two beams were of orthogonal polarisation, there would be no interference and the intensity would be nothing other than the sum of their individual intensities.

It is in the blue term of equation (6.5) that the magic happens. This term is what makes the intensity have a value above or below the sum of the intensities of the individual beams, which is a really strange result. In effect, it is possible to add two light beams and get an intensity less or more than their sum!

6.3 Applications of interference, holography

6.3.1 Example 1: measuring height variation across a sample

For this example, we use a modified MI, shown in figure 6.5, that is fibre-based, rather than a free space one. We explain how interference can be used to measure a surface profile, by using a very simple sample having a step of height h, shown as an inset in figure 6.5.

The intensity at the detector when the beam is at position 1 of the sample is

$$I_1 = I_0[1 + V\cos(\delta_1)] \tag{6.9}$$

and at position 2

$$I_2 = I_0[1 + V\cos(\delta_2)]. \tag{6.10}$$

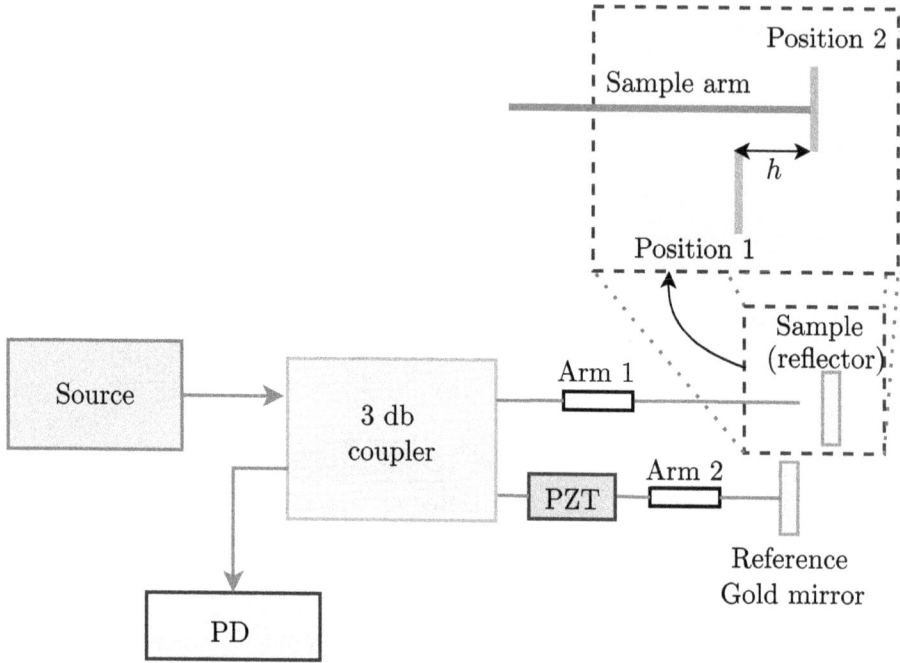

Figure 6.5. Fibre-based Michelson interferometer to measure height difference across a sample.

The phase in each case is

$$\delta_i = \frac{2\pi}{\lambda} \text{OPD}_i, \tag{6.11}$$

where OPD_i is the optical path length difference between the sample and reference beams when the sample beam is at position i. If there was some way of obtaining the phase δ_i, the optical path difference (OPD) could be calculated and the height h of the step would be given by

$$h = \text{OPD}_1 - \text{OPD}_2, \tag{6.12}$$

assuming the sample was in air.

The difficulty in retrieving δ_i from equations (6.9) and (6.10), is that only the measured intensity I_i is known. The other three parameters, namely I_0, V and δ, are unknown. We are not that interested in the first two but δ contains the desired information. How can we extract it without knowledge of I_0 and V? The retrieval of phase from an interference equation has been a topic of much interest for decades now. This is because a wealth of information lies in the phase of a beam reflected from or transmitted through a sample. Take figure 6.6 for example. Figure 6.6(a) shows the intensity image of a blood smear sample taken with a conventional digital microscope. Since the cell is mostly transparent, there is not enough contrast to get an idea of what the cell looks like. However, using a phase extraction technique from an interference pattern of this cell would result in the image shown in figure 6.6(b).

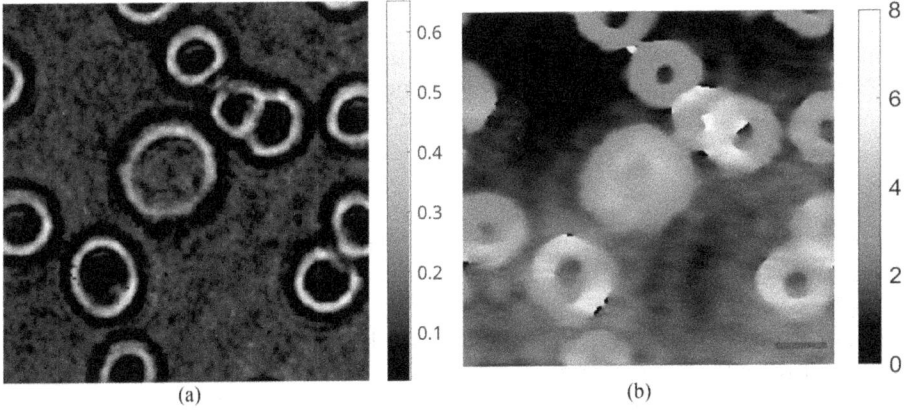

Figure 6.6. Blood smear sample, including a white blood cell in the centre, with some red blood cells around. (a) Intensity image and (b) phase image. The blue scale bar is 5 μm. The white blood cell is clearer in the phase image. Images reproduced from BC's PhD thesis [4], with permission.

6.3.2 Retrieving phase: the four-step method

Let us take a little diversion here to understand how we can retrieve the phase from the expression below:

$$I(x, y) = I_0[1 + V \cos(\delta(x, y))]$$

$$= I_0 + I_0 V \cos(\delta(x, y)). \tag{6.13}$$

The trick is to add a series of known phase differences $\phi(t)$ to one of the interfering beams, such that the interference equation becomes

$$= I_0 + I_0 V \cos[\delta(x, y) + \phi(t)]. \tag{6.14}$$

If the four values of $\phi(t)$ chosen are 0, $\pi/2$, π and $3\pi/2$, respectively, the resulting four interference equations would be

$$I_1 = I_0 + I_0 V \cos[\delta(x, y)] \tag{6.15}$$

$$I_2 = I_0 - I_0 V \sin[\delta(x, y)] \tag{6.16}$$

$$I_3 = I_0 - I_0 V \cos[\delta(x, y)] \tag{6.17}$$

$$I_4 = I_0 + I_0 V \sin[\delta(x, y)]. \tag{6.18}$$

By calculating

$$\frac{I_4(x, y) - I_2(x, y)}{I_1(x, y) - I_3(x, y)} = \tan[\delta(x, y)], \tag{6.19}$$

the phase of the interference equation can be retrieved using only intensity measurements!

> **Point to ponder**: We have already talked about the fact that equations (6.9) and (6.10) have three unknowns in them. Therefore, in principle, the desired unknown δ could be obtained from just three equations. Why do we use four equations, as just discussed?

Different algorithms (using three or more interferograms) exist. They differ in their mathematical properties, especially in their sensitivity to factors such as errors of the phase shifter, noise, etc. All phase-shifting techniques require the capture of a minimum of three interferograms, each having a known added phase. It is important that the entire system stays constant while the interferograms are being captured and that the added phase has no error in it. Because of the extra complexity in the experiment to ensure this, alternative single-interferogram phase retrieval methods have also been explored, a comparison of which is available in [5].

6.3.3 Example 2: the Newton interferometer

In the previous example, interference was used to quantify phase and calculate the height of a step. Interference can also be used to obtain qualitative information about a surface. One common tool found in any optical workshop is the Newton interferometer, shown in figure 6.7.

The lower slab in this figure represents a reference flat, whose surfaces are perfectly flat. The element under test (EUT) is placed on top of the reference flat. Light from the source above is incident on both the EUT and the reference flat. Reflections occur from the top and bottom of the EUT and the top of the reference surface. The air gap between the lower surface of the EUT and the upper surface of the reference flat is very small and the interference between these beams is what is observed as fringes by the operator. If the EUT surface is also flat but there is a wedge, as shown in figure 6.8, then straightline fringes appear. The concept of interference fringes is easy to understand if one thinks of them as a locus of points. Just like isobars are drawn on maps indicating places of equal atmospheric pressure, interference fringes are lines connecting spatial locations with the same optical path length difference (with respect to the reference beam).

A Newton interferometer is often used to check whether there are defects or issues while grinding optical elements such as reference flats, simple plano-convex lenses, etc. The shape, smoothness and orientation of fringes provide a quick analysis of the quality of the surface at that stage of fabrication. If the fringes are perfectly straight parallel lines, this is indicative that the surface of the EUT is also mostly perfectly flat but may have a bump on one end causing an air gap, between it and the reference surface. This kind of defect results in a gap of increasing thickness. The wedge angle α can be quantitatively figured out by the number of fringes visible. Remember that the fringes are formed between the beams reflected from the bottom surface of the EUT and the top of the reference flat. The former is a reflection in a denser medium (the reflection happens at a glass–air interface), which is called

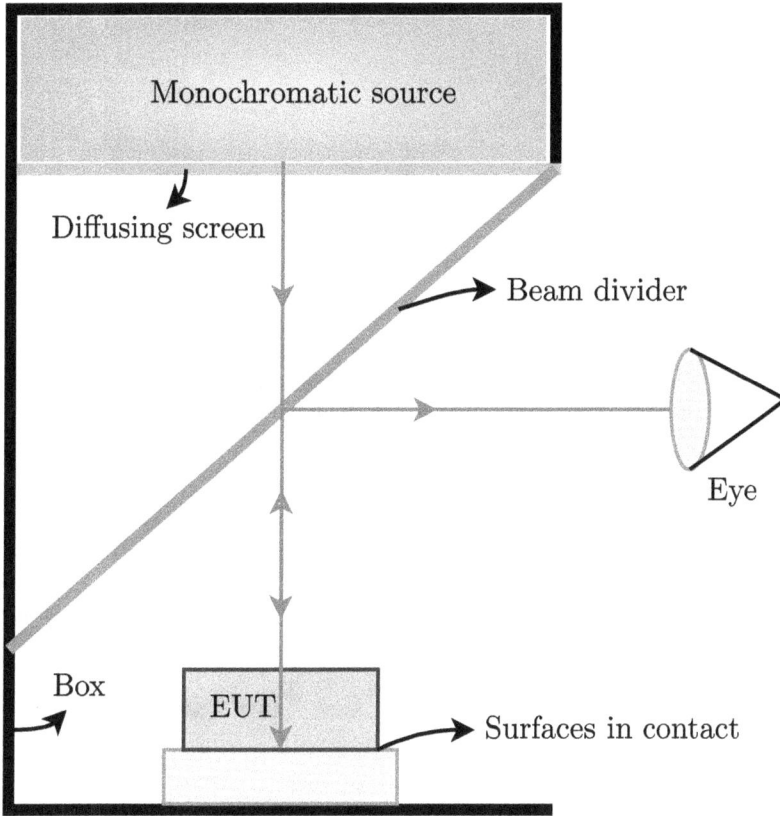

Figure 6.7. Newton interferometer.

internal reflection, and the latter occurs at an air–glass interface, and is called external reflection. Taking into account that a phase change of π occurs when external reflection takes place, a dark fringe is formed whenever the phase difference at position x is an odd multiple of π:

$$2\alpha x + \lambda/2 = (2m + 1)\frac{\lambda}{2}, \qquad (6.20)$$

where m is an integer. This can be simplified to

$$2\alpha x = m\lambda. \qquad (6.21)$$

Bright fringes on the other hand are formed when

$$2\alpha x + \lambda/2 = m\lambda. \qquad (6.22)$$

The distance d, therefore, between any two consecutive dark fringes at locations x_1 and x_2, respectively, can be arrived at from

$$2\alpha(x_1 - x_2) = (m + 1)\lambda - m\lambda.$$

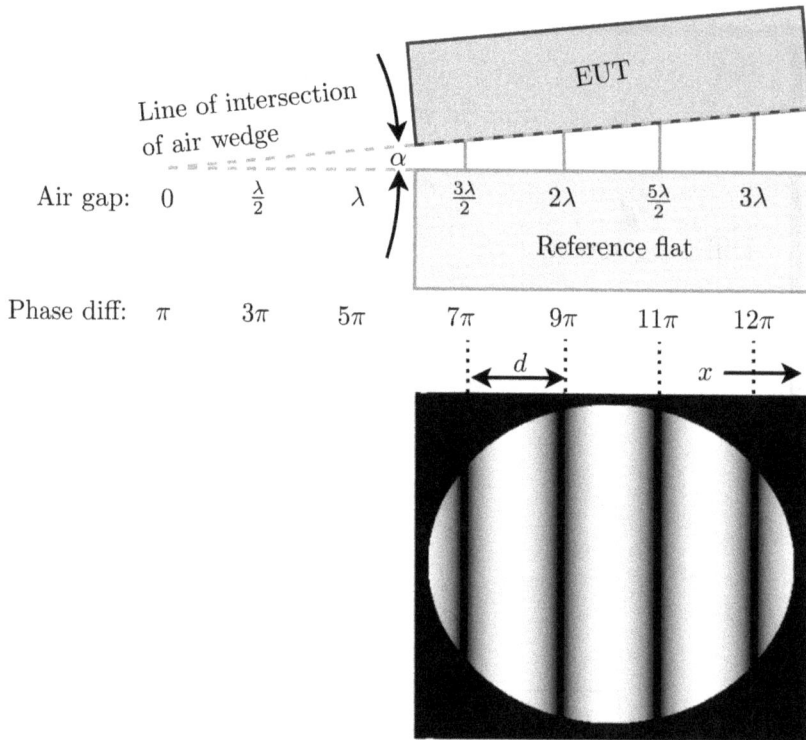

Figure 6.8. Newton interferometer fringes with a wedge between EUT and reference flat.

Therefore,

$$d = (x_1 - x_2) = \frac{\lambda}{2\alpha}. \qquad (6.23)$$

It can also be verified that this is the distance between any two bright fringes as well. The usefulness of this interferometer should be clearer when looking at the fringe patterns of figure 6.9.

In these figures we see a surface that should have been flat but has a defect in the middle. In one case shown in figure 6.9(a), the surface has a bump along the horizontal central line, whereas in figure 6.9(b) there is a hollow area there. Such information would immediately tell the optical engineer how to correct the grinding of the piece to ensure flatness. The value of Δ should ideally go to 0, indicating that flatness has been achieved.

The shape of the fringes depends, of course, on the shape of the EUT and variations are not only caused by errors and defects. If the EUT has a convex or concave lower surface, as shown in figure 6.10(a), the fringes will look like rings as seem in part (b) of the figure.

The distance (or sag) between the elements is now given by

$$\text{sag} = \frac{x^2}{2R}. \qquad (6.24)$$

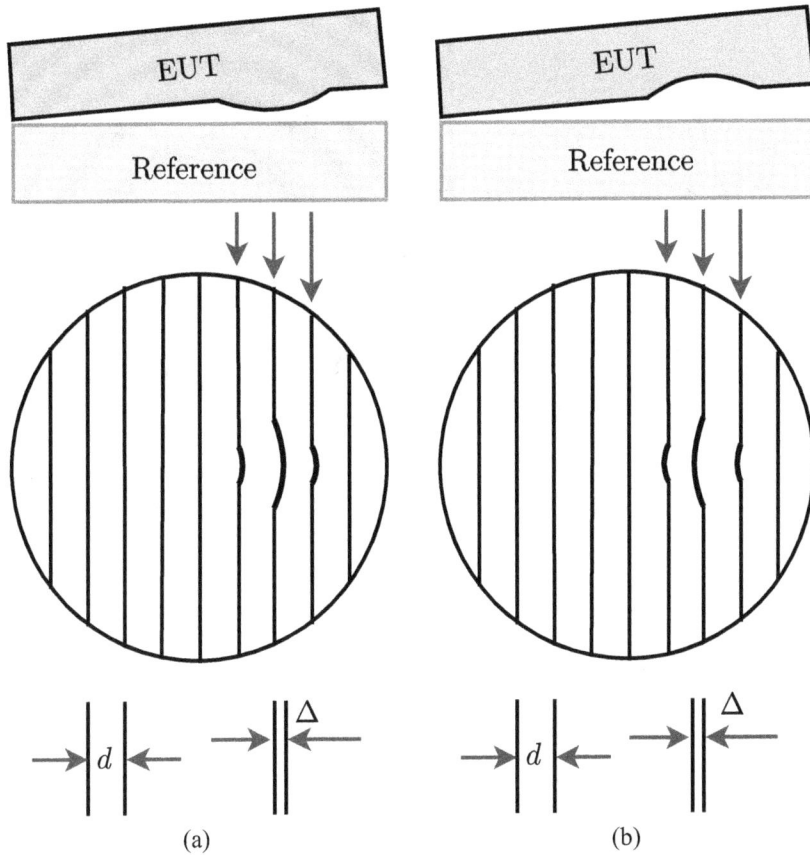

Figure 6.9. Interference image of a surface with (a) a bump and (b) a hole on it.

Since the reflected light travels twice through this distance and the external reflection experiences a π phase shift, the full OPD, at a dark fringe, is given by equation

$$\frac{x^2}{R} + \frac{\lambda}{2} = (2m+1)\frac{\lambda}{2} \qquad (6.25)$$

or

$$\frac{x^2}{R} = m\lambda. \qquad (6.26)$$

The distance of the mth dark fringe from the centre is

$$x = \sqrt{m\lambda R}. \qquad (6.27)$$

Similarly, the constructive interference equation for this system is

$$\frac{x^2}{R} = (2m+1)\lambda. \qquad (6.28)$$

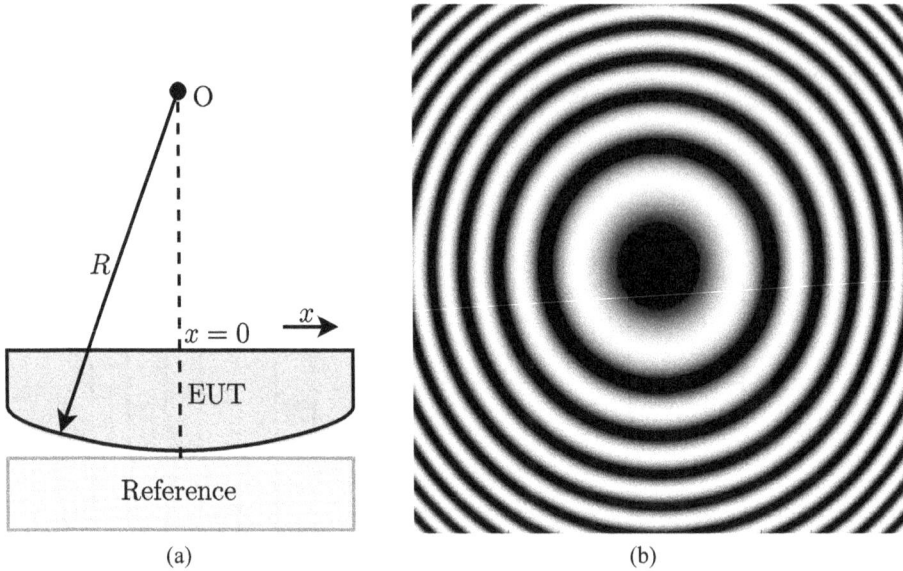

Figure 6.10. (a) Spherical EUT and (b) resulting fringe pattern.

Note that a derivation might result in a term $(2m - 1)$ on the rhs but to keep the term x^2 positive for positive R, we write this as $(2m + 1)$. Both terms simply ensure that the integer is odd.

Usually, bright fringes are used in measurements and therefore the radius of curvature of the EUT from equation (6.28) is

$$R = \frac{x^2}{(2m + 1)\lambda},$$
(6.29)

where x is the distance of the mth bright ring from the centre.

Newton's interferometer provides quick information about surface shape as can be seen in the fringes formed by different surfaces in figure 6.11.

6.4 Thin film interference

When path length differences between interfering beams are in the order of the wavelength of the incident white light, interference presents as colourful patterns that are not only beautiful but also full of information.

Figures 6.12(a) and (b) are photographs of thin films. The former is a common sight on roads due to oil patches created by vehicles. The latter is a photograph of soap bubbles that contain a multitude of colours. The walls of the bubbles are of varying thickness and since each wavelength satisfies the condition for constructive interference only for a particular thickness, the result are these bubbles of ephemeral art.

Since constructive interference occurs for a particular thickness, one can exploit this to make thin films that are reflective or anti-reflective for a particular wavelength. Take the thin film, of refractive index n_f, shown in figure 6.13.

Type of surface reflection occurs from	No tilt between beams	with tilt
Plane		
Spherical		
Cylindrical		
Astigmatic (one type)		

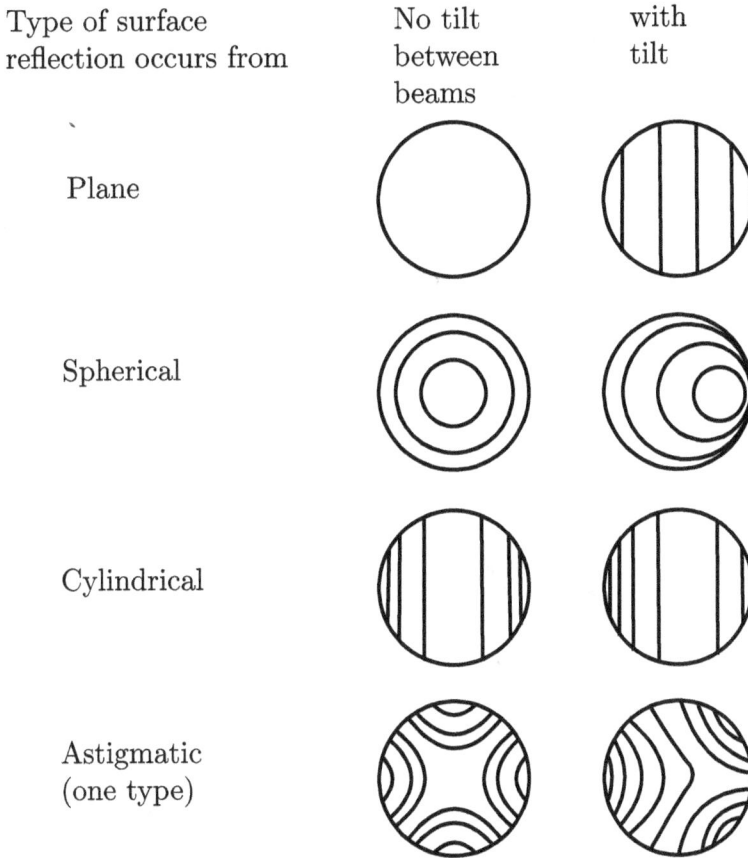

Figure 6.11. Newton interferometer fringes for different surfaces.

In the schematic, light is incident at an angle but to make analysis simpler, we consider the case when the incident angle is $\theta_i = 0$. Part of the incident wave reflects and a part transmits. The latter travels the distance d twice (on the onward and return part) and then exits the film again. These two beams interfere. Assuming that $n_a < n_f < n_s$, the condition for constructive interference is

$$2n_f d = m\lambda, \tag{6.30}$$

where m is an integer.

Similarly, the condition for no reflection (destructive interference) would be

$$2n_f d = (2m + 1)\lambda/2. \tag{6.31}$$

The thinnest film that will ensure zero reflection is approximately quarter-wavelength thick or to be precise is $d = \lambda/(4n_f)$. For students familiar with impedance matching in transformer lines [6], this relation should look familiar. Anti-reflective thin films are commonly used on optics to maximise the amount of light reaching their target by minimising reflection. Such considerations affect our daily lives. For example, we get asked what tint we want our spectacles to be when

(a)

(b)

Figure 6.12. (a) Fringes formed on an oil film on the road. (b) Colourful patterns formed because of the thin walls of a soap bubble. Photo credit: SB.

visiting an optician. This has a direct connection to anti-reflection coatings. This might seem odd as spectacles with an anti-reflection coating on them, should not have any tint (after-all the reflection was cancelled out). However, as can be seen from equation (6.31), this equation works for one wavelength. What happens to the other wavelengths? They will reflect and transmit partially. The resulting colour of the reflected light will be a combination all the wavelengths that were reflected. Typical anti-reflection layers consist of a stack of thin films [7]. By careful choice of materials and thicknesses, the stack can be made anti-reflective over a broad range of wavelengths. However, there is usually some residual reflection and hence, the opticians question about the tint of the spectacles.

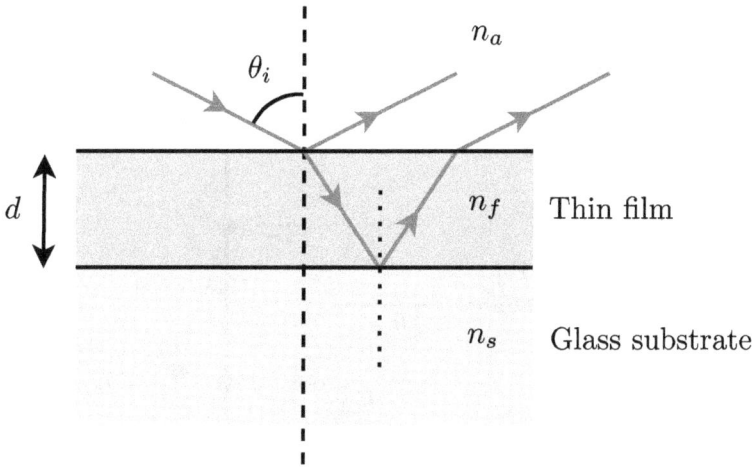

Figure 6.13. Thin film interference.

By now, it should be clear that the phase or optical path length difference has to satisfy a very specific value for destructive interference to take place. What may not be so obvious is that this phase condition is not sufficient to ensure perfect cancellation of two waves. If the two interfering beams are perfectly out of phase, then the condition $\vec{E}_1 - \vec{E}_2$ is achieved but this only means the fields are being subtracted. For perfect cancellation, it is also important that the amplitudes of the fields are equal, i.e.

$$|\vec{E}_1| = |\vec{E}_2|. \tag{6.32}$$

This puts a further constraint on the choice of film, as the condition in equation (6.32) will be satisfied only if $n_f = \sqrt{n_a n_s}$, which can be proved using the Fresnel equations [2].

6.5 Fabry–Pérot interferometer

There are many different types of interferometers. The Fabry–Pérot (FP) interferometer uses multiple beam interference. Such interferometers have many uses, such as in resolving wavelengths in a spectrometer. Because of the multiple beams at play, it is harder to achieve the condition for constructive interference, resulting in much narrower or sharper fringes. In a spectrometer, this results in the improvement of its resolution. A schematic of an FP interferometer is shown in figure 6.14. It consists of a set of highly reflective parallel plates. In the figure, the thickness of the plates is ignored and only three beams are shown emerging from it. In reality, a large number of beams will emerge and be collected by a lens.

Let E_0 be the amplitude of the incident beam. To understand how the FP works, we look first at the emerging beams E_1 and E_2. The path length difference between them is given by

$$\Delta \text{OPL} = n(\text{P}_1\text{P}_2 + \text{P}_2\text{P}_3) - n\text{P}_1\text{Q}. \tag{6.33}$$

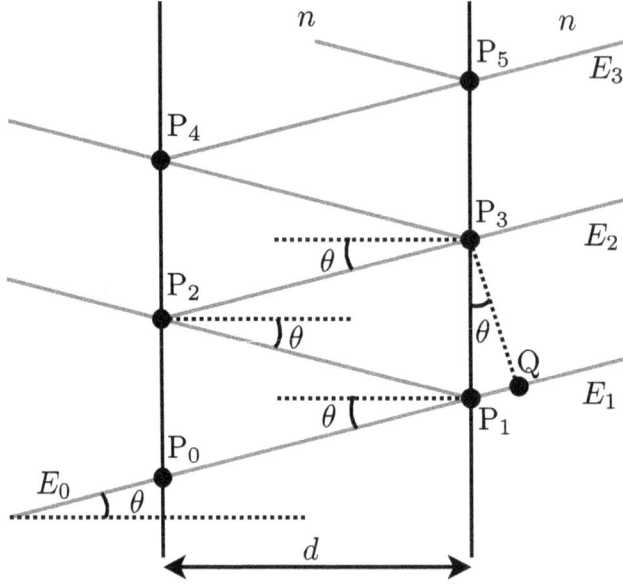

Figure 6.14. Fabry–Pérot interferometer.

However,

$$P_1P_2 = \frac{d}{\cos \theta} \tag{6.34}$$

and

$$P_1Q = P_1P_3 \sin \theta. \tag{6.35}$$

Therefore,

$$\Delta OPL = 2nd \cos \theta. \tag{6.36}$$

This analysis can be extended to find the path length difference for E_i, where $i = 3, 4, 5, \dots$. We also know that a bright fringe will result whenever $2nd \cos \theta = m\lambda$. What can we say about the amplitudes of the interfering beams? Let us assume that the reflectivity of both plates of the FP are identical and equal to r, whereas the transmissivity of the two plates are t and t', respectively. The first beam to emerge from the FP interferometer would have been transmitted through both plates and therefore, the resulting field is

$$E_1 = E_0 tt' \exp jwt - \phi_{P_0P_1}, \tag{6.37}$$

where $\phi_{P_0P_1}$ is the phase acquired as the beam travels from P_0 to P_1. The next beam has travelled further and has reflected on both plates, so its field is

$$E_2 = E_0 r^2 tt' \exp jwt - \phi_{P_0P_1} - \phi, \tag{6.38}$$

where ϕ is the phase because of travelling the distance P_1P_2 and P_2P_3. Each subsequent beam travels a further multiple of this distance. And since every beam acquires the phase $\phi_{P_0P_1}$, it can be left out of all equations, as we are interested in the phase difference between the beams.

The total field (when N beams interfere) at the output of the FP interferometer is

$$E_{\text{total}} = E_0 tt' e^{jwt} + E_0 r^2 tt' e^{j(wt-\phi)} + E_0 r^4 tt' e^{j(wt-2\phi)} + \cdots + E_0 r^{2(N-1)} tt' e^{j(wt-(N-1)\phi)}. \quad (6.39)$$

Recognising this as a geometric progression, it can be rewritten as

$$E_{\text{total}} = \frac{E_0 tt' e^{jwt}}{1 - r^2 \exp(-j\phi)}. \quad (6.40)$$

The intensity of the transmitted beam is $E_{\text{total}} E_{\text{total}}^*$ or

$$I_{\text{total}} = \frac{I_0 (tt')^2}{(1 + r^4) - 2r^2 \cos \phi}. \quad (6.41)$$

In keeping with the conservation of energy, $r^2 + tt' = 1$, the transmitted intensity can be expressed as

$$I_{\text{total}} = \frac{I_0}{1 + F \sin^2(\phi/2)}, \quad (6.42)$$

where F or the coefficient of finesse of the interferometer is given by

$$F = \frac{4r^2}{(1 - r)^2}. \quad (6.43)$$

The transmittance of the interferometer is

$$T = I_{\text{total}}/I_0 = \frac{1}{1 + F \sin^2(\phi/2)}. \quad (6.44)$$

The fringes of the interferometer can be studied using the phase difference, which can be obtained from the OPD given in equation (6.36)

$$\phi = \frac{2\pi}{\lambda} 2nd \cos \theta. \quad (6.45)$$

For the case of normal incidence, this reduces further to

$$\phi = \frac{2\pi}{\lambda} 2nd. \quad (6.46)$$

Constructive interference occurs whenever ϕ is an integral multiple of 2π or $2nd = m\lambda$. We study the FP interferometer initially, assuming a thickness d between the plates and plot the transmittance T versus phase. The latter varies with λ in equation (6.46).

Figure 6.15 shows this variation for two different plate reflectivities. It can be seen that the higher reflectivity results in a much sharper peak. In both cases, the transmittance is high at multiples of ϕ. Closer observation of these phase values will indicate that they occur when the gap between plates is an integral multiple of a particular wavelength.

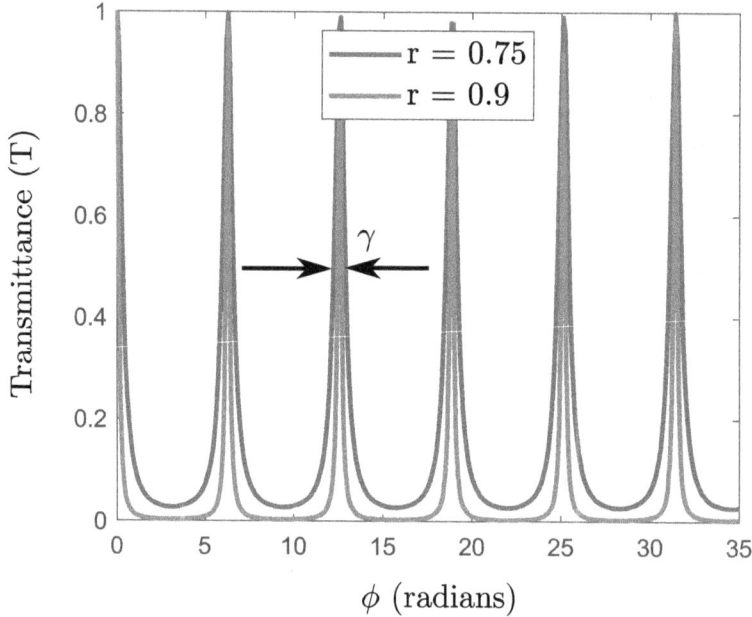

Figure 6.15. Fabry–Pérot transmittance versus phase. The full width at half maximum value is denoted by γ. Data generated by SB.

6.6 Laser Interferometer Gravitational Wave Observatory (LIGO)

The LIGO is a Michelson-interferometer-based system used to detect the presence of gravitational waves (GWs). A schematic is shown in figure 6.16.

If a gravitational wave is incident on the MI, while travelling in the $-z$ direction, space itself is affected, with one arm of the interferometer being stretched and the other compressed. The length of the arms of the interferometer is set up such that no light is detected at the photodetector, i.e. the path lengths ensure perfect destructive interference. However, when a GW crosses the interferometer, the arm lengths are altered and the path length condition is destroyed causing light to appear at the photodetector (PD), indicating the presence of the GW. The interferometer will respond to the GW only if the mirrors are under the condition of free-fall. In other words, the effect of gravity has to be nullified. This is done by suspending the mirrors (or test masses as they are called in LIGO) on a pendulum connected to a seismic isolation stack. Pendula work like mechanical filters. Any disturbance at their resonance frequency gets amplified but frequencies much higher than this do not affect the test mass. The pendulum length is chosen to be in the order of tens of centimeters, so that the resonant frequency is in the order of $f_0 = 1$ Hz [8]. The test mass itself will be unaffected by GWs that have frequencies several orders of magnitude higher than this. It might therefore seem easy to sense the presence of a GW with a MI. However, the strain or fractional change in length of each interferometer arm can be as small as $\epsilon = 10^{-21}$ for a GW of frequency $f_g = 100 \gg f_0$ Hz and wavelength $\lambda_g = 3000$ km. If each MI arm was 1 m long, then the corresponding change would only be 10^{-21} m! To understand how LIGO

Fabry Perot cavities

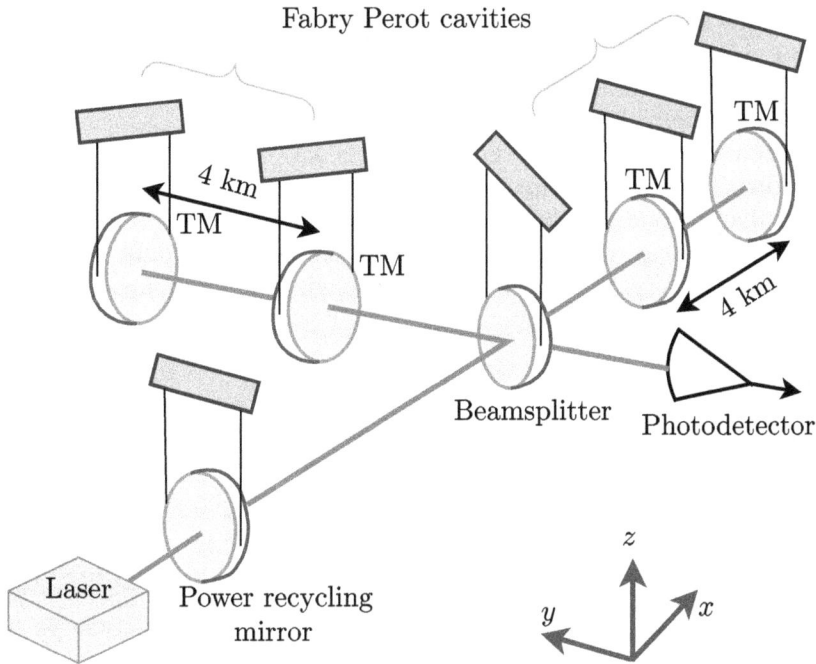

Figure 6.16. LIGO interferometer.

overcomes this problem, let us consider the phase change caused by the GW as it passes through the interferometer:

$$\Delta\phi = \frac{4\pi}{\lambda}\epsilon L \qquad (6.47)$$

$$= \frac{4\pi}{\lambda}\Delta L, \qquad (6.48)$$

where L is the length of one arm and ΔL is the change in that length due to the GW. Given that ϵ is very small, L has to be large for $\Delta\phi$ to be measurable. What might not be obvious is that L cannot be too large. This is because beyond a certain length, if the light is still travelling within the interferometer, the GW may change sign and the effect of $\Delta\phi$ will be cancelled. This means that although a GW did pass through the interferometer, no change will be observed. While the optimal length has been shown to be in the order of $\lambda_g/4$, which for the current example is 750 km; practical LIGO systems have shorter lengths than this. For example, the one at Hanford, WA, has arms of length 4 km, which are much longer than the average Michelson interferometer! To improve the sensitivity to the small phase change, the interferometer is modified so that light circulates N more times within it, with the help of Fabry–Pérot mirrors, as seen in figure 6.16. The phase change then becomes $\Delta\phi = (4N\pi/\lambda)\Delta L$. Because the light bounces back and forth a number of times within the cavity ($N \approx 200$), the light is considered *stored* within the interferometer.

The length of 4 km is long enough that the curvature of the Earth causes the beam to be about 1 m lower at the end of the arm, compared to the starting point! The base of the interferometer is levelled to compensate for this curvature. The 4 km length ensures a path length change in the order of 10^{-18} m. This tiny change was detected in 2015 and in February 2016, an announcement was made by the LIGO and Virgo collaborations, of the first direct observation of gravitational waves. One must remember that many different effects could cause similar or even larger changes in length, e.g. even nearby traffic or changes in temperature would affect the path length. LIGO uses a vast number of complex engineering techniques to isolate the interferometer from these noise sources.

6.7 Holography

Holography is a technique for capturing all the information from an object [9]. By this we mean capture both the amplitude and phase information. Conventional cameras and imaging systems create images that are based only on intensity variations across the object. The lack of phase information renders 2D images. Also, if the object is mostly transparent, e.g. such as plant or animal cells, most of the information lies in the phase rather than the intensity variation, making such objects hard to visualise. The way that holography captures the phase of an object is to create an interference pattern between a wave that either transmits through or reflects from the object with a known reference wave. This first part of the imaging process is called the recording stage. In figure 6.17, an off-axis recording technique is shown. There are other geometries, such as the in-line, the Fourier transform method, or those that yield rainbow and reflection holograms [10]. In the schematic shown, a monochromatic light source of wavelength λ is used to create the interference.

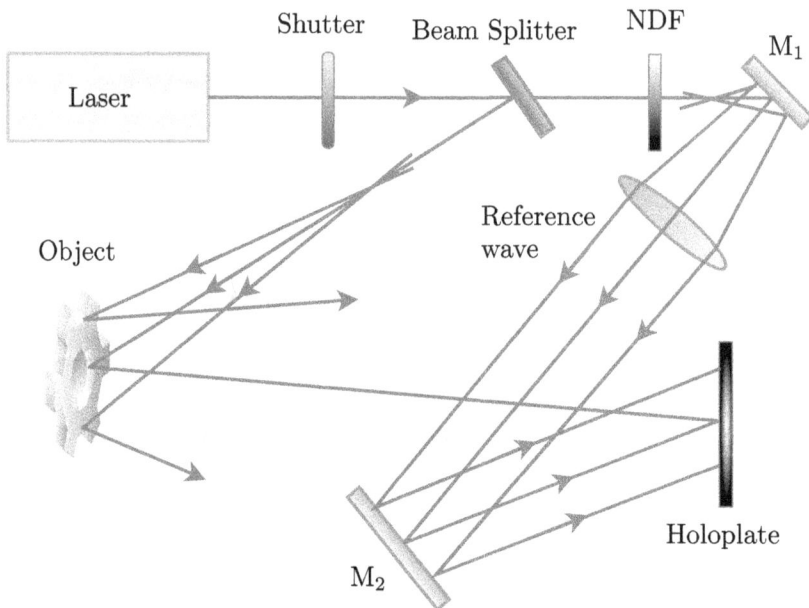

Figure 6.17. Holography recording.

The object beam $O(x, y)$ is unknown (and contains the information of interest). It is represented as

$$O(x, y) = O_0(x, y)\exp[j\phi(x, y)]. \tag{6.49}$$

The reference beam is typically a tilted plane wave. If we assume it is tilted at θ_0 with respect to the y-axis, then the spatial frequency f_R associated with it is $\sin \theta / \lambda$ and the reference wave is given by

$$R(x, y) = R_0 \exp[j2\pi f_R y]. \tag{6.50}$$

A holoplate or a medium sensitive to light is kept at the recording plane. The amplitude incident on the plate is the sum of the object and reference beams or

$$A(x, y) = O(x, y) + R(x, y). \tag{6.51}$$

The intensity is

$$I(x, y) = O_0^2(x, y) + R_0^2 + 2O_0(x, y)R_0 \cos[2\pi f_R y - \phi(x, y)]. \tag{6.52}$$

The holoplate is exposed to this intensity for a time T_e and the exposure is $E(x, y) = I(x, y)T_e$. The plate is developed leaving behind an element with transmittance

$$T(x, y) = t_0 - \beta \, [I(x, y) - R_0^2]T_e, \tag{6.53}$$

where t_0 is a dc reduction in intensity as the light passes through the plate. β is a constant for the particular holoplate that indicates how it responds to the incident intensity. The developed plate is called a hologram.

The interference pattern will be a spatial function of the phase variation across the object. It is possible to create an image from the interference pattern, in the second 'reconstruction' part of the imaging process, shown in figure 6.18.

In the reconstruction process, the hologram is illuminated by only the reference wave. In other words, the following operation is taking place:

$$
\begin{aligned}
T(x, y)R(x, y) &= t_0 R(x, y) \\
&\quad - \beta T_e R(x, y)\{O_0^2(x, y) + 2O_0(x, y)R_0 \cos[2\pi f_R y - \phi(x, y)]\}
\end{aligned} \tag{6.54}
$$

$$
\begin{aligned}
&= t_0 R(x, y) - \beta T_e R(x, y)O_0^2(x, y) \\
&\quad - \beta T_e R_0^2 O_0^*(x, y) \exp [\,j(4\pi f_R y - \phi(x, y))] \\
&\quad - \beta T_e R_0^2 O_0(x, y) \exp [\,j(\phi(x, y))].
\end{aligned} \tag{6.55}
$$

There are four terms in equation (6.55). Let us look at each one of them to understand what happens when a hologram is reconstructed. Each term represents a beam that is generated when the reference beam is incident on the hologram. The first term $t_0 R(x, y)$ is nothing other than the reference wave, with its amplitude altered. The original reference wave had amplitude R_0 and now the wave has amplitude $t_0 R_0$. There is another wave travelling in the same direction as the incident reference wave. It is of the form $-\beta T_e R(x, y)O_0^2(x, y)$. The terms of interest are $-\beta T_e R_0^2 O$ and $-\beta T_e R^2(x, y)O^*(x, y)$. The former is the object wave (with a slightly different amplitude from the original wave), and the latter is the complex conjugate

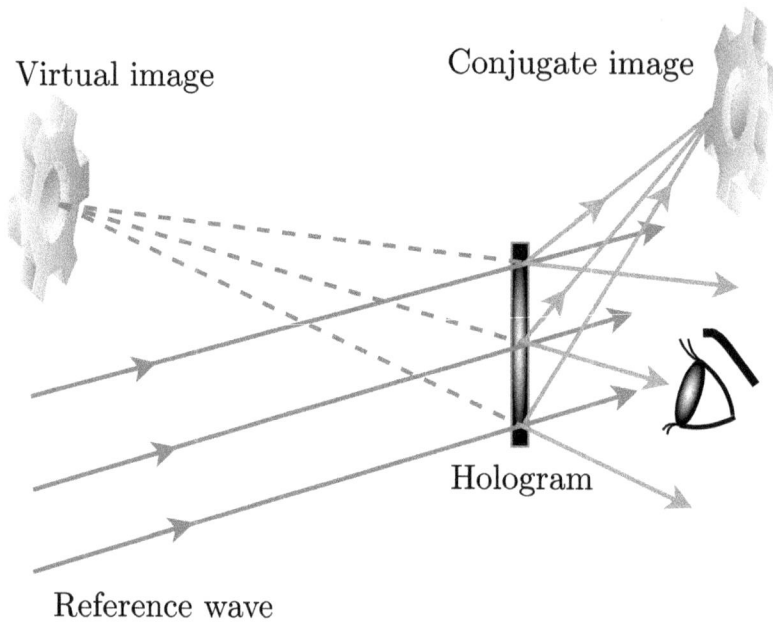

Figure 6.18. Holography reconstruction.

object wave. It is this wave that creates a real image of the object with amplitude and depth information (courtesy of the phase of the wave). One could place a camera or screen at the location where this image is formed and capture a 2D version of it. Alternatively, the 3D version can be viewed in real time. If one were to look through the hologram as seen in figure 6.18, it will appear as if there is an image there. This is actually a virtual 3D image that cannot be captured on a camera at that location. Holograms can be designed to be visible in white light. Such holograms have many uses, e.g. they are present on the currency notes of many countries and product labels, as they are harder to copy than simple images. Holography has several very important applications that include:

- Security (to provide product authenticity).
- Holographic interferometry (for non-destructive testing).
- Holographic microscopy (for imaging of phase objects).
- Data storage.
- Art.

6.7.1 Rainbow holography

Conventional holography requires the use of coherent light. However, the holograms on currency notes or credit cards are visible under regular white light. Several differences in the recording set-up are required to achieve this. First, a slit is placed in front of the object to improve the spatial coherence of the light being used. Also, a lens is used to image the object onto the holoplate. On reconstruction, the image forms at the hologram itself, which reduces the coherence requirements on the source.

When a rainbow hologram is reconstructed, two images are formed, one of the object but also that of the slit. To the viewer, it will appear as if they are viewing the object through the slit. From any one location, only a small horizontal section of the image can be seen. To see another part of the object, the viewer needs to tilt their head and change the viewing angle. However, at each angle, a different wavelength reconstructs another part of the image, resulting in an object seen, with varying colour (in a direction perpendicular to the slit length). This is why these types of holograms are called rainbow holograms.

6.7.2 Holographic interferometry

This technique uses the interference between two beams, one of which was generated from a hologram, for non-destructive testing or metrology [11, 12]. There are three main holographic interferometry (HI) techniques, namely:

- *Double exposure*
  Both exposures are recorded sequentially. The object may be subjected to a load during the second exposure. The hologram is a permanent record of the effect of that load on the object.
- *Real time*
  The hologram is recorded, processed and then replaced in the original system. Reconstruction is carried out and the reconstructed object interferes with the real object wave (the object can be loaded in real time) highlighting any differences between them.
- *Time average*
  This technique is useful when studying vibrating objects. The recording is carried out with exposure times $\gg$ period of vibration.

HI can be used in a wide variety of applications. For example, it could be used to measure temperature, density or velocity distributions in a wind tunnel or it is often used in the aeronautical industry to help locate defects in airplane wings. In the latter example, the basic idea of HI would be to record an initial hologram of airplane wings, without any strain applied. Before developing the hologram, another recording of the same object but under pressure would then be recorded. On reconstruction, this doubly exposed hologram would result in a 3D image of the object overlaid by a fringe pattern indicating the deformation experienced by the object due to the applied pressure. A defect would show up in the nature of the fringes formed.

6.8 Moiré interferometry

Moiré interferometry is a technique that converts two sets of small-period fringes into a pattern with a larger period. There are a number of ways this is beneficial. For example, any image can be thought to be made up of a number of spatial frequencies (just like any electrical signal can be decomposed into a number of sinusoidal frequencies). The spatial frequencies may be too small to be resolved. However, by overlaying the spatial frequencies of an image with 1D grating image of similar frequency, but at a slight tilt to the original grating, it is clear that another set of fringes form. These are the Moiré fringes. A simulation demonstrating this is shown in figure 6.19. Part (b) is the same

Horizontal 1D grating

d_g

(a)

Same grating with a slight tilt

d_g

(b)

Resulting fringes when
grating overlap each other

d

(c)

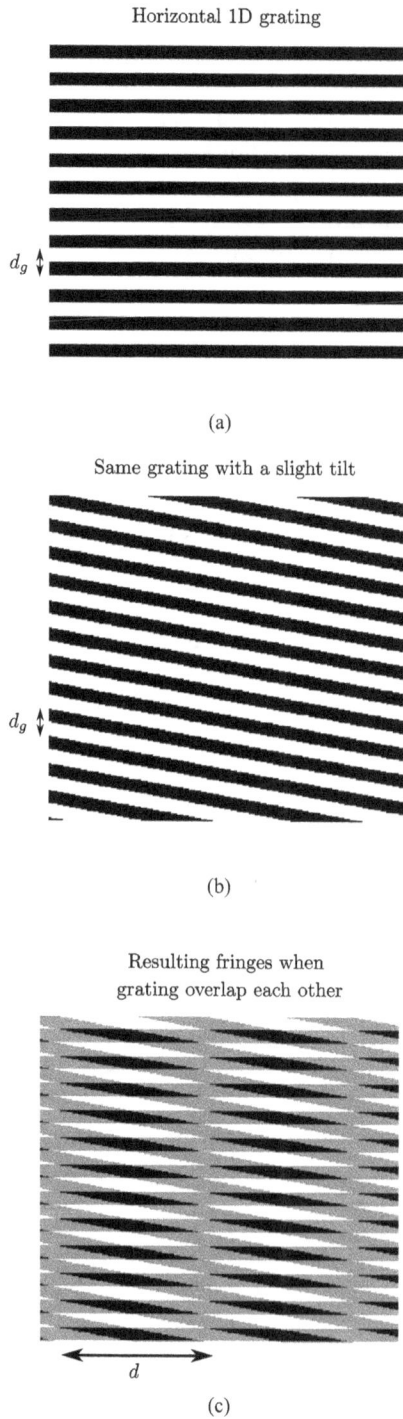

Figure 6.19. (a) 1D grating with period d_g. (b) The same grating as shown in (a) but slightly tilted. (c) Moiré fringes when the two gratings overlap. A new fringe frequency d is visible. Data generated by SB.

(a)   (b)

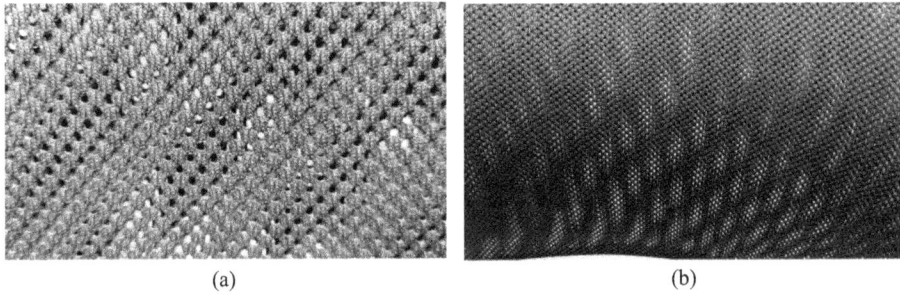

Figure 6.20. (a) Single layer of mesh of the backrest of (my) office chair. (b) Photograph of the same chair with Moiré fringes visible, as both mesh layers are in focus. Photo credit: SB.

grating as (a) but with a tilt, i.e. both gratings have period d_g. Overlaying the two gratings gives rise to the pattern in (c), a new set of fringes with period $d > d_g$. These fringes are called Moiré fringes. Because their period is larger, they are easier to resolve and provide information about the original spatial frequency $1/d_g$.

Moiré fringes are often seen in unexpected places. For example, the backrest of an office chair sometimes has two overlapping meshed layers, an example of which is seen in figure 6.20(a)

It can be shown [13] that the period of the Moiré fringes is given by

$$d = \frac{d_{g1}d_{g2}}{\sqrt{d_{g1}^2 + d_{g2}^2 - 2d_{g1}d_{g2}\cos\theta}}, \tag{6.56}$$

where d_{g1} and d_{g1} are the periods of each of the gratings and θ is the tilt of one grating with respect to the other. The tilt of the Moiré fringes is ϕ, where

$$\tan\phi = \frac{-d_{g1}\sin\theta}{d_{g2} - d_{g1}\cos\theta}. \tag{6.57}$$

6.9 Problems

1. A 532 nm diode laser has a bandwidth 2 nm. What is its coherence length?
2. A laser output of 650 nm is fed into a Mach–Zehnder interferometer and the number of fringe shifts (bright to dark or dark to bright) on gradually filling a square gas chamber kept on one of the interfering arms with a gas of unknown refractive index is noted to be 32. The side length of the gas chamber is 1 cm. Estimate the refractive index of the gas.
3. A glass substrate having an index 1.5 is to be coated with a silicon nitride film to eliminate the reflection of normally incident light (wavelength $\lambda_0 = 1550$ nm), as shown in figure 6.21. The refractive index of silicon nitride at this wavelength is 1.99. Derive an equation to calculate the minimum thickness of the film so that it acts as an anti-reflection coating. If you could pick a film with a specific refractive index, what value would be best? Explain the reason.

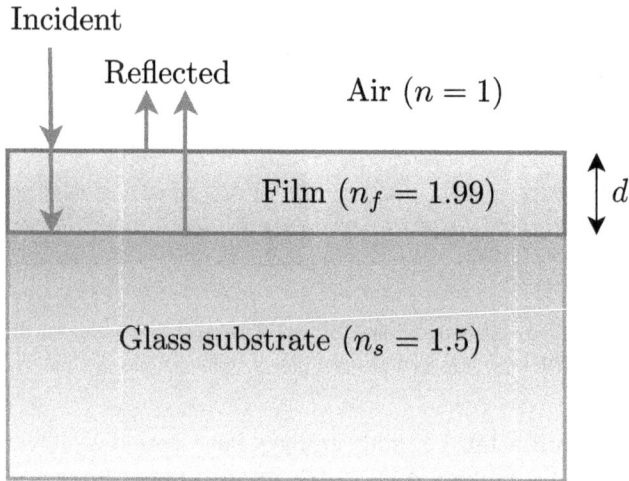

Figure 6.21. Calculation of thin film thickness for an anti-reflection coating.

4. The reflectivity of the mirrors of an FP interferometer or etalon is given by $r = 0.92$. What is the full width half maximum (FWHM) of each transmission peak?

5. Derive equations (6.56) and (6.57). Hint: Consider 1D amplitude gratings and assume that one of them is parallel to either the x- or y-axis. For the former, the grating lines will be given as $y = md_{g1}$, where m is an integer. Write a similar equation for the tilted grating in terms of its period and the index n.

6. A 1D periodic pattern has a period d, such that $d/2 < p$, where p is the size of the pixel of the camera being used to image it. A grating of period 3 μm is tilted at $5°$ and the Moiré fringes are observed. If $p = 1.2$ μm, will the fringes be visible on the camera?

References

[1] Deng Y and Chu D 2017 Coherence properties of different light sources and their effect on the image sharpness and speckle of holographic displays *Sci. Rep.* **7** 5893

[2] Hecht E 2012 *Optics* (New York: Pearson)

[3] Michelson A A and Morley E W 1887 On the relative motion of the Earth and the luminiferous ether *Am. J. Sci.* **34** 333

[4] Chandraprasad B T 2025 Methods for optical phase retrieval from a single off-axis interference pattern *PhD Thesis* Indian Institute of Technology Madras

[5] Chandraprasad B T, Vayalamkuzhi P and Bhattacharya S 2021 Transform-based phase retrieval techniques from a single off-axis interferogram *Appl. Opt.* **60** 5523–33

[6] Shevgaonkar R K 2005 Electromagnetic waves *Electrical and Electromagnetic Waves* (New York: McGraw-Hill)

[7] Macleod H A 2001 *Thin-Film Optical Filters* (Electrical and Electronic Engineering Series) (Boca Raton, FL: CRC Press)

[8] Physics of LIGO, Lecture 2 https://dcc.ligo.org/public/0033/G000164/000/G000164-00.pdf. (Accessed: 17 June 2023)

[9] Gabor D 1948 A new microscopic principle *Nature* **15** 4098

[10] Hariharan P 1996 *Optical Holography: Principles, Techniques and Applications* 2nd edn (Cambridge: Cambridge University Press)

[11] Heflinger L O, Wuerker R F and Brooks R E 1966 Holographic interferometry *J. Appl. Phys.* **37** 642–9

[12] Petrov V, Pogoda A, Sementin V, Sevryugin A, Shalymov E, Venediktov D and Venediktov V 2022 Advances in digital holographic interferometry *J. Imaging* **8** 196

[13] Saveljev V, Kim S and Kim J 2018 Moiré effect in displays: a tutorial *Opt. Eng.* **57** 030803

IOP Publishing

Introduction to Ray, Wave, and Beam Optics with Applications

Shanti Bhattacharya

Chapter 7

Diffraction and diffractive optics

While it may not be obvious, the effects of diffraction impact our daily life in many different ways. Let us take the simple case of a single aperture, as shown in figure 7.1 (a). If a plane wave is incident on the slit and the size of the slit $b \gg \lambda$, geometric optics predicts that the patch of light on a screen, after the slit, will be of the same size as the slit. However, if b is small compared to λ, then it appears as if light has *leaked* in to the shadow region, as seen in figure 7.1(b). Closer observation will show that not only is there light in the supposed shadow region but that it has a pattern consisting of bright and dark regions.

How did this happen? To answer this, we use the Huygens principle [1], which states that each point on a wavefront can be considered to be a secondary source. The secondary sources on a plane wave are shown in figure 7.2(a). The wavelets generated from these secondary sources are shown in the same figure. The envelope of these, shown in red, is nothing other than the next wavefront.

In figure 7.2(b) the plane wavefront is incident on a slit or opening. The Huygens principle is very useful here, as it allows us to visualise the wave after the slit. In the middle of the wavefront (i.e. the regions B and C), the envelope looks like a plane wave but at the ends, the envelope is curved. As the wave travels even further from the slit, the wave spreads more into what should have been the shadow region. The presence of light in this area can only be explained using wave theory. This spreading of light as it travels in free space is called diffraction. One might think that it occurs only if light goes through an aperture but the extremities of any finite-sized beam will themselves act like an aperture and cause diffraction. Huygens principle raises as many questions as it answers. For example, why do we draw the envelope or tangent to the wavefront only on one side? Why does not the wave travel backward? What is the physical basis of the secondary sources? As with many other theorems, under certain circumstances and assumptions, they yield fairly accurate results and given their simplicity are powerful tools for understanding physical phenomena. A more rigorous explanation of diffraction was developed over the 300 years following

doi:10.1088/978-0-7503-5497-4ch7

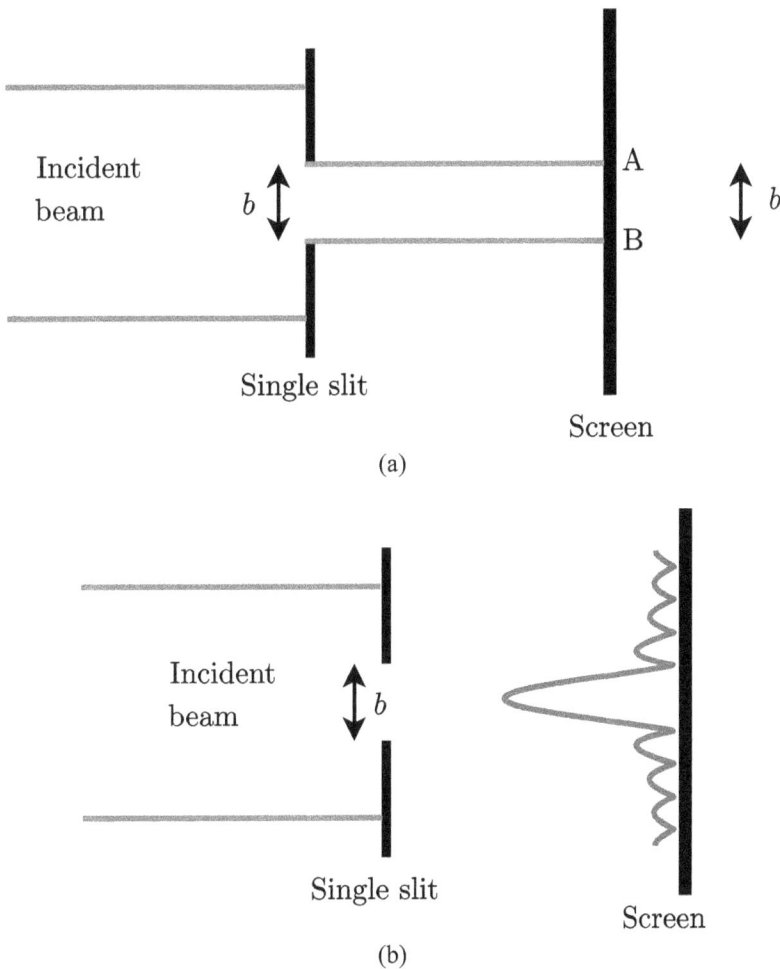

Figure 7.1. Light travels through a slit (a) with no diffraction. (b) A cross section of the intensity at the screen when diffraction is taken into account.

Huygens that answered these and other questions. Some of the key scientists who contributed to this are Young, Fresnel, Poisson, Maxwell and Kirchhoff. More recently, David Miller proposed a theory that addresses some of the inaccuracies of the original Huygens theory [2].

We already know that the new wavefront is created from the superposition of the wavelets from each secondary source. Interference is also the result of the super-position of beams. In fact, diffraction is nothing other than a form of interference. The latter term is used when a small number of beams are interacting with each other. For example, two beams in an interferometer. Since those beams go through apertures and have finite sizes, they experience diffraction as well. However, interferometers are set up such that the predominant effect observed is due to the superposition of the two interfering beams. When considering light travelling

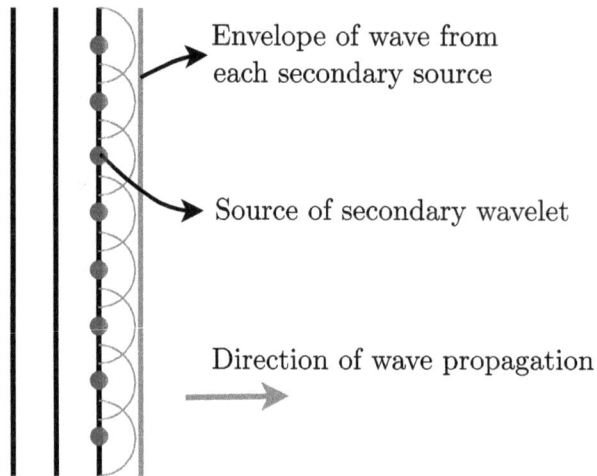

(a)

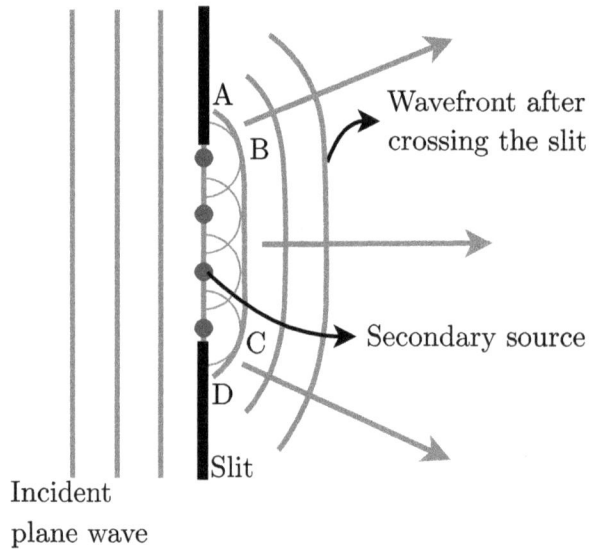

(b)

Figure 7.2. (a) Demonstration of Huygens principle with secondary sources and (b) waves diffracting or spreading out after travelling through a slit.

through a single aperture, one can consider that there are an infinite number of beams (from all the secondary sources) participating in the process. However, if the aperture size is much larger than the wavelength of the light travelling through the system, diffraction effects happen over such a small region that they become insignificant. In any case, in the centre of beam, far from the aperture edges, the effect of that superposition results in a beam similar to the original. However, at the edges, the story is quite different. For small apertures, the effects of diffraction can no longer be ignored. At this point, it is useful to recollect that light is an electromagnetic wave. When the aperture size $b \gg \lambda$, all the components of the fields can be treated as scalar waves and can be considered independent of each other. However, at the boundaries and when b is closer to or smaller than λ, the vectorial nature of the field cannot be neglected. This means that the various field components may be coupled in these regions, causing the intensity patterns close to the boundary to be quite different from interference in the middle of the wavefront. We refer to this phenomenon as diffraction, rather than interference, although both are caused by the superposition of beams. In figure 7.2(b), we can see the wave that was originally a plane wave spreading out after the aperture. This spread increases as the wave propagates. Clearly, propagation itself affects a wave. Diffraction therefore needs to be taken into account when any beam propagates. Each type of beam behaves in a different way when experiencing diffraction. We have already seen this when studying Gaussian beams and comparing their propagation with respect to plane waves. The perfect non-diffracting beams drawn in textbooks to depict a plane wave do not exist in reality although, under certain conditions, close approximations to them may be achieved.

Diffraction is an important phenomenon with many implications and applications. We have already encountered its negative effects in section 2.8, where we saw how it limits the resolution of an imaging system. Since Goodman's *Introduction to Fourier Optics* [3] was first published in 1968, it has continued to remain the key text for anyone seeking to understand the theory behind diffraction, as well as the mathematics (Fourier techniques) for analysing it. Readers wishing to understand diffraction in more detail are strongly encouraged to read Goodman's book. In this chapter we present some of the key results necessary to analyse beams in systems taking diffraction into account. This will also be useful in understanding the working of some basic diffractive optical elements.

7.1 Diffraction theory

Before we begin this section, we restate the basic premise of diffraction, which is that it occurs when many different beams interact and superimpose. We know that these beams could be generated when a wavefront interacts with the boundaries of an aperture. What is also of importance, but may not be obvious, is that one can design optical elements that create multiple beams such that their superposition results in some desired outcome. In this way, diffraction becomes a tool rather than a hindrance. Diffractive optical elements (DOEs) can manipulate light in non-intuitive and powerful ways. Ironically, a DOE has been used to beat the diffraction limit in microscopy, a discovery that won its inventor, Stefan Hell, the Nobel prize in Chemistry in 2014 [4]! We will delve into this in more detail in later chapters.

A rigorous approach to calculating the diffracted field (i.e. the spatial distribution of the amplitude and phase) as light propagates in a medium or through an optical element requires one to start with Maxwell's equations and derive the scalar wave equation as well as the wave equation near any boundaries, where the scalar components are coupled. If some assumptions are made, however, these derivations

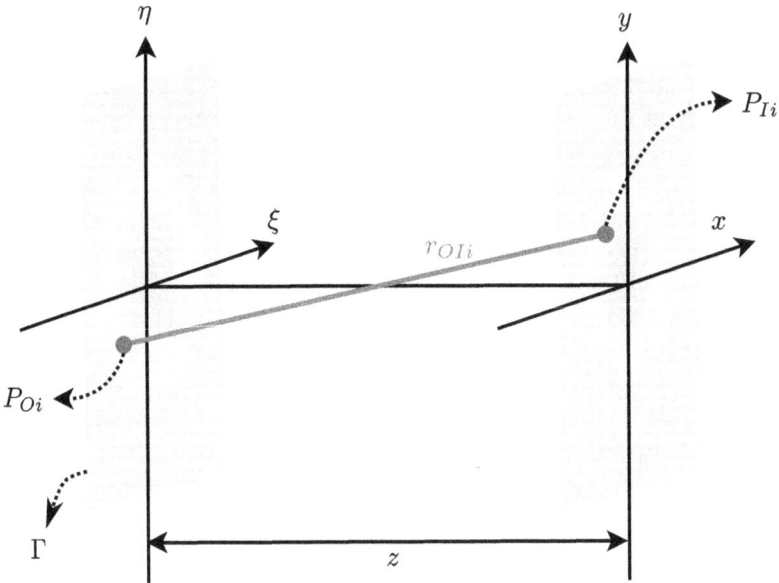

Figure 7.3. The derivation of a diffracted field.

can be made much simpler. In this section, we present a brief description of how the diffracted field $E(x, y)$ can be calculated from the known field $E(\epsilon, \eta)$. More details of the derivation can be obtained from [3]. Figure 7.3 is useful to understand the derivation of the diffracted field. $E(x, y)$ and $E(\epsilon, \eta)$ can be thought of as the output and input planes of interest.

Light from every point P_{Oi} on the aperture indicated by the surface Γ in the object plane contributes to the amplitude at each point P_{Ii} in the image plane. The index i references every point in each plane. The following equation is a mathematical representation of this, where the field amplitude at any point in the image plane can be obtained from

$$E(x, y) = \frac{1}{j\lambda} \iint_{\Gamma} E(\epsilon, \eta) \frac{\exp(jkr_{Oli})}{r_{Oli}} \frac{z}{r_{Oli}} \, d\epsilon \, d\eta. \qquad (7.1)$$

The distance r_{Oli} is the distance from each point in Γ contributing to the amplitude at a particular point P_{Ii}. The double integral in the equation ensures that the effect of all points in the object space is taken into account. This statement is known as the Huygens–Fresnel principle. The main assumption made in arriving at this was that the distance $r_{Oli} \gg \lambda$, which means that the diffracted field cannot be calculated very close to the aperture with this equation. The distance is given by

$$r_{Oli} = \sqrt{z^2 + (x - \epsilon)^2 + (y - \eta)^2}. \qquad (7.2)$$

Depending on the value of r_{Oli} with respect to parameters such as wavelength and aperture sizes, different assumptions can be made with regard to equation (7.2).

Diffraction is therefore usually split into two broad regions, one called near-field (or Fresnel diffraction) and, not surprisingly, the other called far-field (or Fraunhofer diffraction).

7.1.1 Fresnel diffraction

In Fresnel diffraction, we are interested in the field at distances close to the input plane, satisfying the condition $z > x, y$. Therefore, equation (7.2) can be approximated to

$$r_{0li} = z + \frac{(x - \epsilon)^2 + (y - \eta)^2}{2z}. \tag{7.3}$$

Substituting this in relevant places in equation (7.1), results in

$$E(x, y) = \frac{e^{jkz}}{j\lambda z} \iint E(\epsilon, \eta) \exp\left(j\frac{k}{2z}\left[(x - \epsilon)^2 + (y - \eta)^2\right]\right) d\epsilon\, d\eta. \tag{7.4}$$

There are different ways of interpreting this equation. By gathering certain terms together, the equation has the form

$$E(x, y) = \frac{\exp(jkz)}{j\lambda z} \exp\left[j\frac{k}{2z}x^2 + y^2\right] \times$$
$$\iint_{-\infty}^{\infty}\left[E(\epsilon, \eta)\exp\left[j\frac{k}{2z}\epsilon^2 + \eta^2\right]\right]\exp\left[-j\frac{2\pi}{\lambda z}x\epsilon + y\eta\right]d\epsilon\, d\eta. \tag{7.5}$$

This equation can be seen to be similar to a Fourier transform operation apart from the exponential term in the square bracket of the kernel of the integral. In the equation, this appears in blue.

Equation (7.4) can also be expressed as a convolution equation of the form

$$E(x, y) = \iint E(\epsilon, \eta)h(x - \epsilon, y - \eta)\mathrm{d})x\mathrm{d}y, \tag{7.6}$$

where

$$h(\epsilon, \eta) = \frac{e^{jkz}}{j\lambda z} \exp\left(j\frac{k}{2z}[\epsilon^2 + \eta^2]\right) \tag{7.7}$$

is the convolution kernel or in other words, a Fresnel diffraction impulse response [3]. Borrowing ideas from signal processing [5], its Fourier transform can then be considered a transfer function for Fresnel diffraction given by

$$H(f_x, f_y) = e^{jkz} \exp\left[-j\pi\lambda z\left(f_\epsilon^2 + f_\eta^2\right)\right]. \tag{7.8}$$

Using the convolution theorem, equation (7.4) can be written as

$$\boxed{E(x, y) = \mathrm{DFT}^{-1}\left[\mathrm{DFT}\left[E(\epsilon, \eta)\right]H\left(f_\epsilon, f_\eta\right)\right].} \tag{7.9}$$

Both equations (7.5) and (7.9) can be used to evaluate the near-field diffraction pattern. The latter is called the angular spectrum method. Although both are derived from the same equation, differences arise in the output field during practical implementation. This will be explored in section 7.6.

7.1.2 Fraunhofer diffraction

When the distance from the aperture increases, i.e. $z \gg x, y$, equation (7.2) can be approximated to

$$r_{0Ii} = z + \frac{x^2 + y^2 - 2x\epsilon - 2y\eta}{2z}. \tag{7.10}$$

This is called the Fraunhofer approximation, and substituting this in relevant places in equation (7.1) results in

$$
\begin{aligned}
E(x, y) = \frac{\exp(jkz)}{j\lambda z} \exp\left[j\frac{k}{2z}(x^2 + y^2) \right] \\
\times \iint_{-\infty}^{\infty} E(\epsilon, \eta)\exp\left[-j\frac{2\pi}{\lambda z}(x\epsilon + y\eta) \right] d\epsilon\, d\eta.
\end{aligned}
\tag{7.11}
$$

The integral of equation (7.11) is nothing but the Fourier transform (FT) of $E(\epsilon, \eta)$. The implication of this result is that just by propagation through a large distance, an optical beam undergoes a FT operation. It can also be shown that the FT operation can be carried out by the use of a lens, as the field after a lens of focal length f is given by

$$E(x, y) \propto \iint_{-\infty}^{\infty} E(\epsilon, \eta)\exp\left[-j\frac{2\pi}{\lambda f}(x\epsilon + y\eta) \right] d\epsilon\, d\eta. \tag{7.12}$$

The plane of observation in this case is at a distance corresponding to the focal length of the lens. It is interesting to note the similarity between this equation and equation (5.28), which can be used to obtain the field immediately after a lens.

How does one decide whether Fresnel or Fraunhofer's theory is valid? A rule of thumb is to calculate the Fresnel number F given by

$$F = \frac{a^2}{z\lambda}, \tag{7.13}$$

where a is the radius of the opening light is travelling through and z is the distance of the observation screen from it. If $F \approx 1$, near-field or Fresnel diffraction theory can be employed but if $F \ll 1$, the Fraunhofer theory is more valid.

An alternative guideline is that the Fraunhofer approximation is valid if

$$z > \frac{8a^2}{\lambda}. \tag{7.14}$$

7.2 Diffraction case studies

In the next few sections, we look at specific cases of diffraction.

7.2.1 Diffraction through a single rectangular slit

The experimental set-up shown in figure 7.4 can be used to observe the intensity at some distance from a single slit.

Let us assume that the slit is of width b and the wavefront passing through it is made up of a large number of point sources P_1, P_2, ... , P_n equally spaced by distance Δ. What the figure shows are that all rays, from the secondary sources, travelling in the same direction can be brought to focus at one point by the use of a lens. The figure shows only one set of parallel rays (at an angle θ to the optical axis), but there are similar sets in all directions in the region after the slit. θ can be thought of as the angle of diffraction. The intensity across the plane in the far-field can be calculated by superposing these waves and taking the appropriate phase differences into account. Consider the waves travelling from P_1 and P_2. The phase difference between these neighbouring points is

$$\phi = \frac{2\pi}{\lambda}\Delta \sin \theta. \tag{7.15}$$

Assuming that each point source generates a wave of amplitude A, the resultant field E_Q at point Q on the observation screen due to the n point sources is

$$E_Q = A\left[\cos wt + \cos(wt - \phi) + \cos(wt - 2\phi) + \cdots + \cos(wt - (n - 1)\phi)\right] \tag{7.16}$$

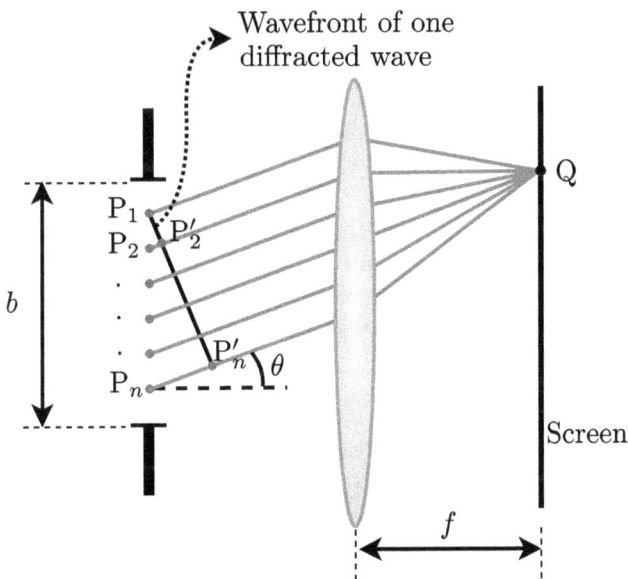

Figure 7.4. Experiment to study diffraction through a single slit.

$$= A \exp(jwt)[1 + e^{-j\phi} + e^{-2j\phi} + \cdots + e^{-j(n-1)\phi}]. \tag{7.17}$$

Given that this is a geometric progression of n terms, the equation simplifies to

$$= A e^{j[wt-(n-1)\phi/2]}\frac{\sin n\phi/2}{\sin \phi/2}, \tag{7.18}$$

which can be written in terms of an amplitude and phase as

$$= E_0 \exp(j\Phi). \tag{7.19}$$

The amplitude of this field at Q is

$$E_0 = A\frac{\sin n\phi/2}{\sin \phi/2}. \tag{7.20}$$

We started this calculation with a finite number of secondary sources n to keep the derivation simple. However, since every point on the wavefront acts as a secondary source, we let $n \to \infty$, which in turn implies that $\Delta \to 0$. The width of the slit is $b = (n - 1)\Delta$, but since n tends to infinity $n\Delta \to b$ is equally valid.

Under these assumptions,

$$E_0 = A_0\frac{\sin \beta}{\beta}, \tag{7.21}$$

where $A_0 = nA$ and $\beta = \pi b \sin \theta/\lambda$.

The field at Q taking the real part of equation (7.18) reduces to

$$E_Q = A_0\frac{\sin \beta}{\beta} \cos(wt - \beta) \tag{7.22}$$

and the intensity is

$$I_Q = I_0\frac{\sin^2\beta}{\beta^2}. \tag{7.23}$$

I_0 represents the intensity on the axis, i.e. $\theta = 0$. By varying θ or in other words β, one can trace the intensity across the observation plane. The location of the maximum and minimum intensity can be found by observing how β varies. For example, the minima will occur whenever $\beta = m\pi$, where m is an integer and $m \neq 0$. This is because $\sin \beta = 0$ in those cases. The diffraction pattern for a single slit is plotted in figure 7.1(b).

7.2.2 Diffraction from a circular aperture

Since many optical elements are circular and have radial symmetry, it is important to understand the diffraction that occurs at such apertures. The diffraction pattern from a circularly symmetric aperture results in what is called an Airy pattern, already mentioned briefly in section 2.8. Here, we present the diffraction intensity equation for an aperture of radius r, which can be seen to be similar to equation (7.23) for a single slit:

$$I_Q = I_0 \frac{4J_1^2(u)}{u^2}, \tag{7.24}$$

where $u = 2\pi/\lambda r \sin\theta$ and $J_1(u)$ is a first-order Bessel function, a function commonly encountered when dealing with circular apertures or functions. $J_1(0) = 0$ but

$$\lim_{u\to 0} \frac{2J_1(u)}{u} = 1.$$

The intensity associated with an Airy pattern will be proportional to

$$\left[\frac{2J_1(u)}{u}\right]^2,$$

with the zeroes appearing at

$$u = \frac{2\pi}{\lambda} r \sin\theta = 3.832,\ 7.016,\ 10.174\ \ldots. \tag{7.25}$$

The far-field diffraction pattern of a circular opening is shown in figure 7.5.

Most lenses are circular and it is clear that they will be affected by diffraction too. Studying the diffraction from a circular convex lens, as shown in figure 7.6, will lead to some important ideas about imaging resolution.

A plane wave travelling along the optical axis is incident on this lens but, due to diffraction, fringes are formed in the far-field (or focal plane). The diameter of the beam incident on the lens is $D = 2r$ and the radius of the first dark ring caused by diffraction is R.

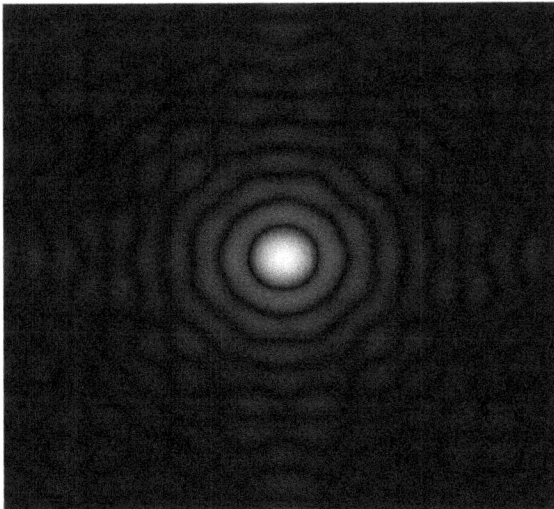

Figure 7.5. Diffraction pattern of a circular opening.

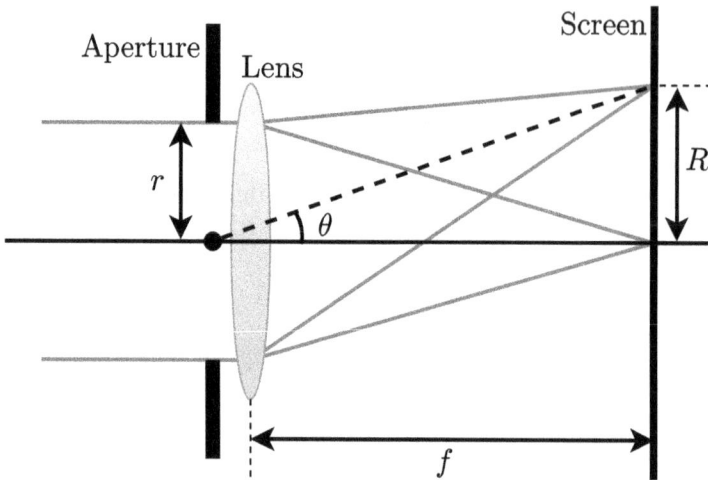

Figure 7.6. Set-up to study diffraction due to a circular lens of focal length f.

From the figure we note that

$$\tan \theta = R/f, \tag{7.26}$$

but we also know from equation (7.25) that the angle of the beam forming the first zero is

$$\sin \theta = \frac{3.832\lambda}{2\pi r}. \tag{7.27}$$

Since θ is small, the equations can be equated and

$$R \approx \frac{f\lambda \, (3.832)}{2\pi r} \tag{7.28}$$

$$= 1.22\lambda f/D. \tag{7.29}$$

Point to ponder: Compare equations (5.48) and (7.29). Both give the actual spot size achievable when focusing a beam with a lens. The former is for a Gaussian beam and the latter for the ideal plane wave, taking diffraction into account. Notice the difference. Is the achievable spot size, then, not just a function of lens parameters and wavelength?

The focused spot will have a number of dark rings around it whose diameters can be calculated by replacing the number in blue of equation (7.28) with the appropriate values from equation (7.25).

7.2.3 Diffraction through a double slit

Imagine that we have two slits as shown in figure 7.7(a) and are interested in the Fraunhofer diffraction pattern at the image plane shown in figure 7.7(b).

Each slit is of width b and the slits are a distance d apart. The field E_1 due to the first slit will be given by equation (7.22). The field E_2 due to the second slit will have an additional phase $\phi = (2\pi/\lambda)d \sin \theta$, due to the displacement of the second slit with respect to the first.

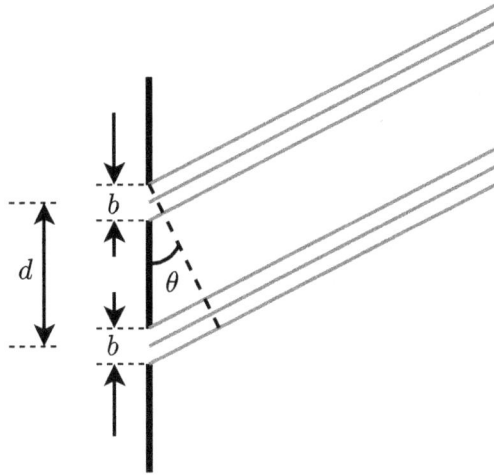

(a)

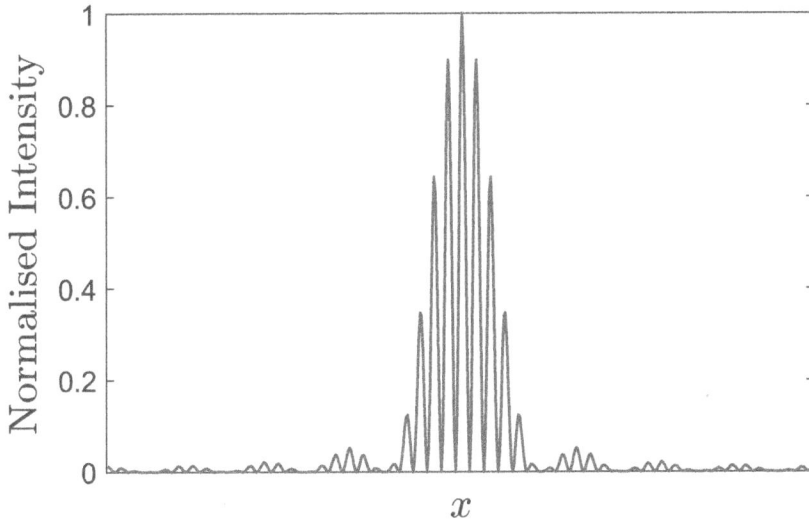

(b)

Figure 7.7. (a) Double slit and (b) intensity pattern in the far-field.

$$E_2 = A_0 \frac{\sin \beta}{\beta} \cos(wt - \beta - \phi). \tag{7.30}$$

Of course, the field at point Q on the image plane will be the vectorial summation, in other words, the *interference* of both fields yielding

$$E_Q = E_1 + E_2. \tag{7.31}$$

With some simple trigonometry, the expression can be reduced to

$$E_Q = A_0 \frac{\sin \beta}{\beta} \cos(\phi/2)\cos(wt - \beta - \phi/2). \tag{7.32}$$

The intensity is then

$$E_Q = 4I_0 \frac{\sin^2 \beta}{\beta^2} \cos^2(\phi/2). \tag{7.33}$$

The intensity across the plane can be obtained by sweeping through the range of θ values. An example of such an intensity pattern is shown in figure 7.7(b). The two-slit system is a treasure trove as far as studying and understanding diffraction and interference. The interference pattern consists of peaks of high intensity surrounded by regions of lower intensity. Where those peaks occur can be tuned, and peaks can even be made to disappear based on the relative sizes of b and d. Again, the choice of those parameters will control which phenomenon is dominant. In equation (7.33), the red part indicates the intensity variation due to diffraction from each slit and the blue, the interference of light between slits.

7.3 The diffraction grating

We already see from figure 7.7(b) that the resulting intensity peaks are narrower than those of a single slit diffraction pattern, as seen in figures 7.1 and 2.16. The peaks become even narrower when the number of slits increases. A diffraction grating can be considered to be a periodic array of a large number of slits. The resulting intensity peaks are called the orders of the grating. The orders can be very narrow and well-separated. Gratings are the workhorse of the spectroscopy industry, in addition to from their use in other applications. In equation (7.33) the intensity of the diffracted field was calculated across the observation plane. With an element like a grating, however, we know that the intensity peaks in the diffracted field correspond to points where constructive interference takes place. To study a grating therefore, we develop the equations that look only at these points.

7.3.1 Grating terminology

- *Amplitude and phase gratings*
   Gratings come in many different forms and flavours. All the examples of diffraction we have studied so far involved an aperture or apertures which allowed light to pass through but blocked it in other places. These were all

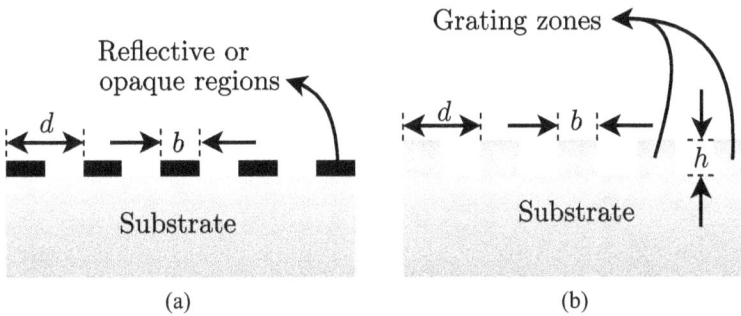

Figure 7.8. (a) Amplitude grating and (b) phase grating.

examples of amplitude elements. Gratings may also be amplitude elements. However, any amplitude element has poor efficiency, as a large fraction of the light does not reach the plane of interest. Practical systems use phase gratings rather than amplitude ones. Instead of periodically blocking light, a periodic phase change is introduced. The two types of gratings are shown in figure 7.8. In both cases, the period of the grating is d. The phase grating in figure 7.8(b) is sometimes called a binary step grating.

- *Transmission and reflection gratings*

Gratings can be used in transmission or reflection. In figure 7.8(a), the periodic structures (each of width b) of the grating lie on a transparent substrate. If the structures are opaque, light passes through the remaining regions of the grating of width $d - b$, and the opening ratio of the grating is given by $(d - b)/d$. Such a grating represents a transmission amplitude grating. If the structures are reflective and the reflected diffraction pattern is observed, the grating is a reflection amplitude grating of opening ratio b/d. These are both amplitude gratings, as only a part of the light plays a role in the desired diffraction. On the other hand, in figure 7.8(b), all of the light is used. Multiple beams are generated because the structures of the grating ensure that different regions of the incident beam travel through different path lengths.

7.3.2 Deriving the grating equation

The amplitude grating shown in figure 7.9 can be used to calculate the equation of a grating. Angles are always measured with respect to the normal to the grating.

In the figure, monochromatic light is incident on the grating with angle θ_i. Diffracted light is observed on the screen in the far-field. Since the grating is a periodic structure, we look at what happens over a single period. We are interested only in the points where constructive interference occurs. At those places, the path length difference between the rays arriving from either end of a period must satisfy

$$\text{OPD} = n(\text{BC} + \text{CD}) = m\lambda, \tag{7.34}$$

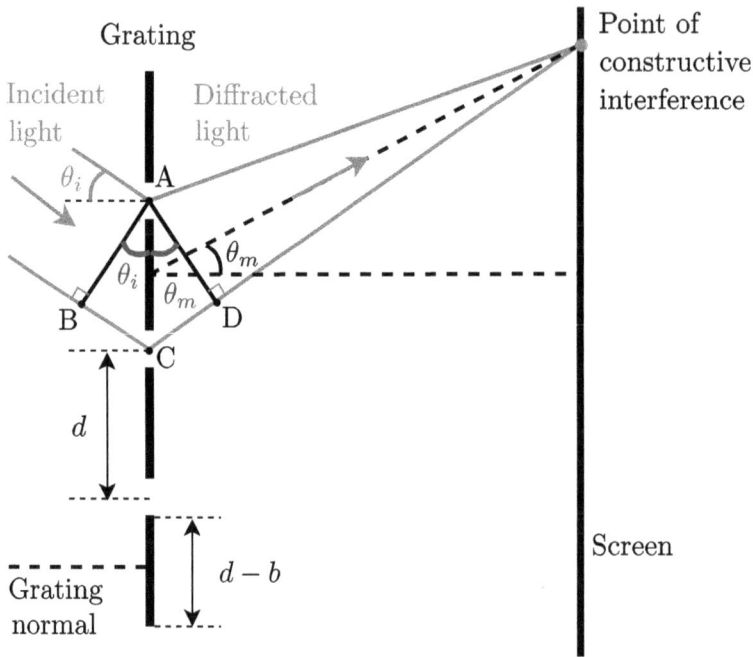

Figure 7.9. Diffraction grating equation.

where n is the refractive index surrounding the grating and m is an integer. The distances BD and BC can be obtained from the triangles ABD and ABC in the figure:

$$\sin \theta_m = \frac{CD}{d}$$

and

$$\sin \theta_i = \frac{BC}{d}.$$

Substituting this into equation (7.34) results in

$$nd(\sin \theta_m - \sin \theta_i) = m\lambda. \tag{7.35}$$

If we assume that the grating is surrounded by air, the equation becomes

$$d(\sin \theta_m - \sin \theta_i) = m\lambda. \tag{7.36}$$

In equation (7.35) the incident angle is taken as negative keeping in mind the sign convention defined in section 2.4. What is the significance of m? Let us assume that only one wavelength is incident on the grating. m can take different integer values, which implies that the beam will be split into orders travelling in specific directions. The sign of the orders of the grating is as indicated in figure 7.10. θ_m represents the angle of diffraction of each order. If the incident beam has more wavelengths

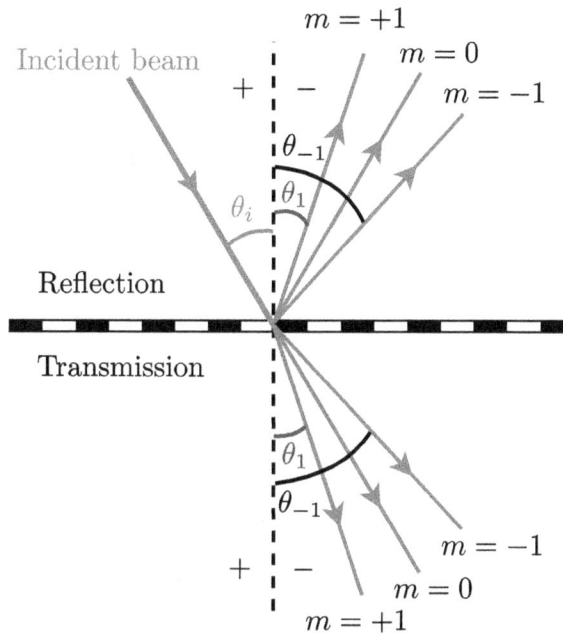

Figure 7.10. Convention of numbering diffraction grating orders.

present, each wavelength generates a set of orders travelling in different directions. This is why gratings are so useful in spectroscopy.

If the beam is incident normal to the grating, the equation reduces to the more familiar grating equation:

$$d \sin \theta_m = m\lambda. \tag{7.37}$$

Point to ponder: Equation (7.36) is valid for phase gratings as well, as it gives us information about the angular positions of the orders of the grating irrespective of the type of grating. What information about the orders is not available in this equation?

7.3.3 What is wrong with binary diffraction gratings?

The grating in figure 7.8(b) is called a binary phase grating, as it has two phase levels associated with it. Such gratings are relatively simple to fabricate, which also lowers their cost. The problem with such gratings is the large number of orders (see figure 7.11) that are generated, effectively reducing the efficiency of the element. In particular, light is often lost to the zeroth order, which can be thought of as the direction refracted or specularly reflected light would take.

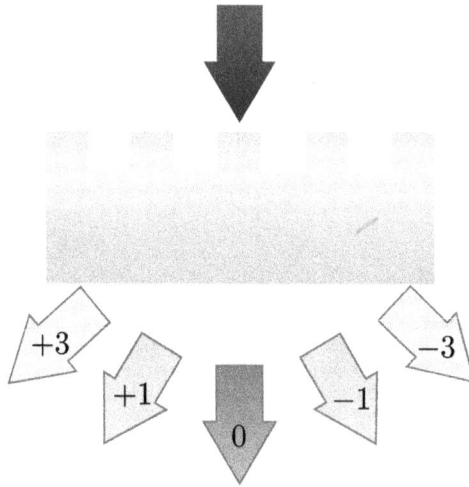

Figure 7.11. Orders of a binary diffraction grating with 50% duty cycle.

7.3.4 An alternative approach to understanding gratings

We know that the peak intensities in the far-field of a grating arise at points where constructive interference takes place. We could have used the Fraunhofer diffraction equation (7.11) to arrive at the same result. This equation is nothing other than the Fourier transform of the incoming field. In this case, the field, which will be the binary phase, looks like a square wave. So the problem reduces to that of finding the FT of a square wave. This example is always studied in any basic textbook on signal processing, as such signals are fundamental in every communication and electronic device. Using the language of signals and system, the FT of a square wave [5] of period $T = 2\pi/w_0$ and duty cycle $2T_1/T$ is given as

$$X(jw) = \sum_{m=-\infty}^{\infty} \frac{2\sin(mw_0 T_1)}{m}\delta(w - mw_0). \tag{7.38}$$

$X(jw)$ represents the complex amplitudes of the frequency components present, which are determined by the argument of the impulse function $\delta(w - mw_0)$. The FT of such a grating shows us that such a wave comprises an infinite number of sine waves of different amplitudes, phases and periods.

It is interesting to note how the amplitudes of each frequency component vary with changes to the duty cycle. Take the case of the 50% duty cycle grating, in which $2T_1/T = 0.5$ or $T_1 = T/4$. Substituting this into the argument of the term in blue in equation (7.38) yields the sine term

$$\sin\left(\frac{m\pi}{2}\right). \tag{7.39}$$

Anytime the integer m is even, equation (7.39) goes to 0. This is why the key contributing terms for such a grating are shown in figure 7.12 with the even orders missing.

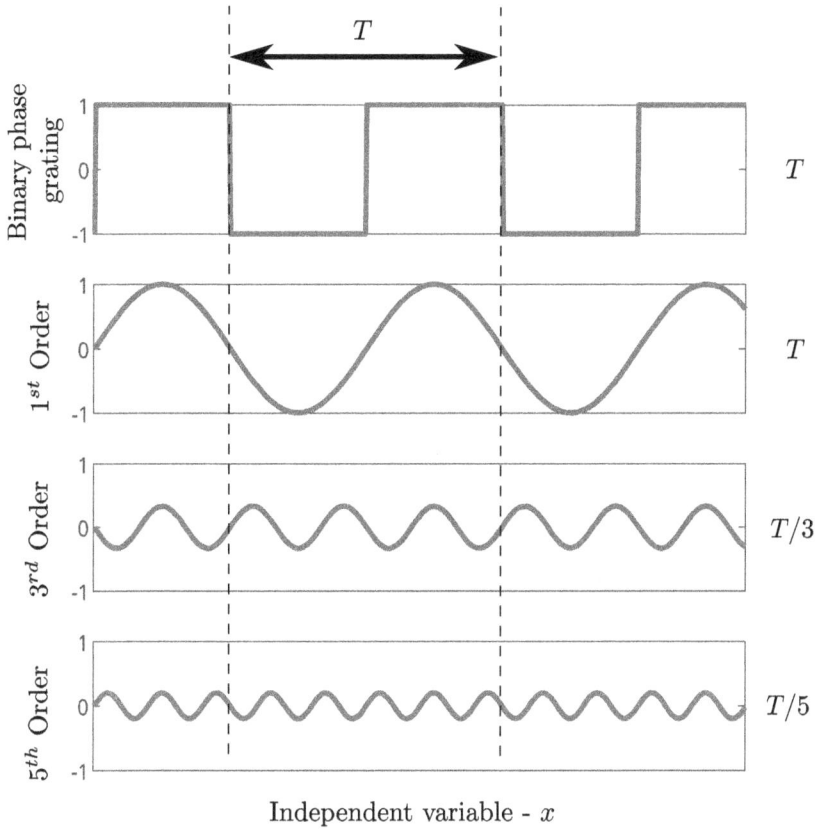

Figure 7.12. FT components of a binary diffraction grating with 50% duty cycle. Data generated by SB.

The first order has the same period as the square grating itself. Although an infinite number of odd orders need to co-exist to create this grating, one can see from their diminishing amplitudes that the higher orders contribute less and less. This was plotted for a 50% duty cycle grating, which resulted in the missing even orders. So, duty cycle greatly affects the orders of a grating but what role does the height of the grating play?

Consider a beam normally incident on the phase transmission grating of figure 7.8. Part of the beam travels through the material of height h and index n, while the remaining part travels through the same distance h but in air. The phase difference between these two parts is

$$\phi_T = \frac{2\pi}{\lambda}h(n-1). \tag{7.40}$$

For the 50% duty cycle transmission grating, if $\phi_T = \pi$, the light travelling directly through will undergo complete destructive interference. This means the zeroth order of the grating will be cancelled out. The height required to achieve this is

$$h_T = \frac{\lambda}{2(n-1)}. \tag{7.41}$$

For a reflection grating, the phase difference is $\phi_R = 2\pi(2h)/\lambda = \pi$, and therefore, the height needed to cancel the zeroth order is

$$h_R = \frac{\lambda}{4}. \tag{7.42}$$

The efficiency of the various orders of a binary phase grating with height given by equation (7.41) is shown in figure 7.13, for gratings with different duty cycles.

In the figure on the top, which relates to the first order, as expected, we see that the efficiency goes to 0 when the duty cycle is 50%. For all the orders, we notice that the efficiency for very small or very large duty cycles is almost 0. This should not be surprising, particularly if we look at figure 7.14.

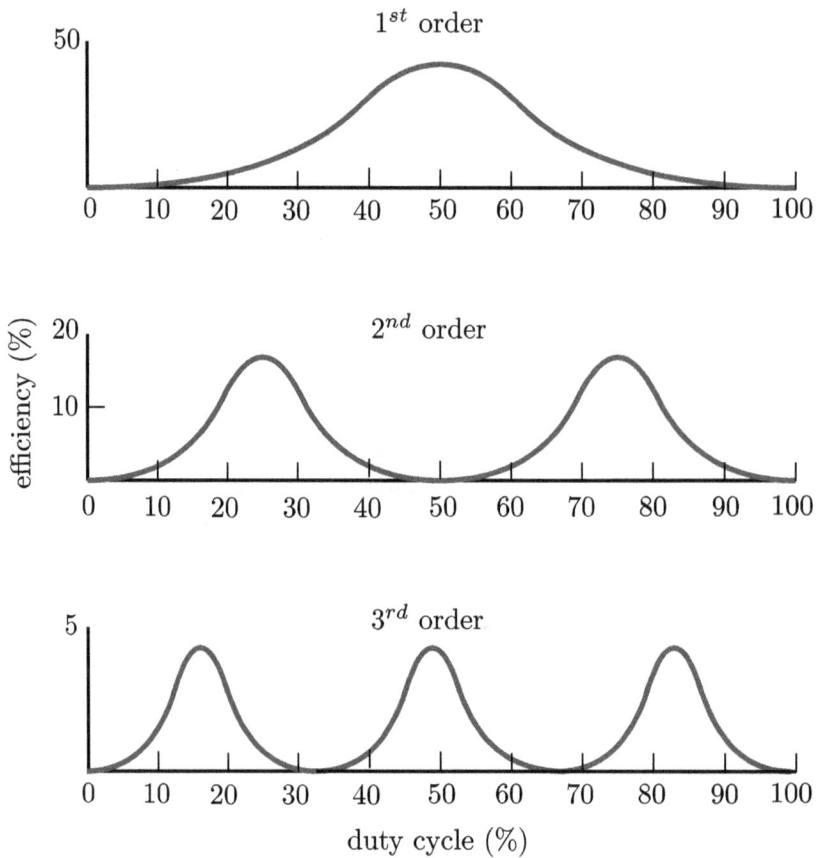

Figure 7.13. Variation in efficiency of the orders of a binary diffraction grating with duty cycle. Data generated by SB.

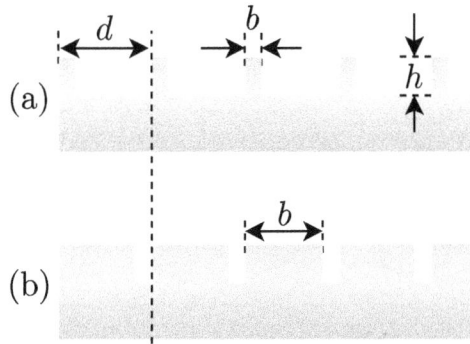

Figure 7.14. Binary gratings with different duty cycles.

If the duty cycle $\to 0$ or 1, as shown in figure 7.14(a) and (b), the element becomes more like a plane surface than a grating, and most of the light will be forced into the zeroth order. The other point to note is that perfect cancellation of the zeroth order only occurs when the duty cycle is 50%. This is because while the height ensures the 180° required for subtraction of the interfering beams, the 50% duty cycle ensures the subtracted terms have equal amplitude.

7.3.5 The blazed diffraction grating

We have seen how the efficiency of a grating is affected by a variety of parameters. In particular, for a step grating, the +1 and −1 orders were always of equal intensity due to the symmetry of the grating. The blazed grating is a form of grating that breaks this symmetry and ensures maximum efficiency in one order. This occurs for one particular wavelength, referred to as the blaze wavelength of the grating. A schematic of such a grating is shown in figure 7.15(a). The parameters of importance are the period d, as well as the angle of the blazed structure, θ_B.

Apart from breaking the symmetry of the step grating, the blazed grating is able to improve efficiency by ensuring that the direction of the specularly reflected light depends on the blaze angle and not on the normal to grating. In figure 7.15(b), the zeroth order ($m = 0$) is shown for light incident at θ_i. Substituting $m = 0$ in equation (7.36) tells us that $\theta_m = \theta_i$. This should not be surprising, as this is what we would expect as specular reflection from a flat surface.

The trick to achieving high efficiency in the blaze grating is to choose the incident angle such that the incident and reflected beams travel along the same path, as shown in figure 7.15(c). This happens when the incidence angle is equal to the blaze angle, i.e. $\theta_i = -\theta_B$. In that case, equation (7.36) reduces to

$$2d \sin \theta_B = m\lambda. \tag{7.43}$$

This particular case is called Littrow configuration [6].

Grating normal

Facet normal

θ_B θ_i θ_B θ_m $0^{th}\, order$

θ_B

d

(a)

θ_i θ_r

θ_B

(b)

$\theta_i = \theta_B$

θ_B

(c)

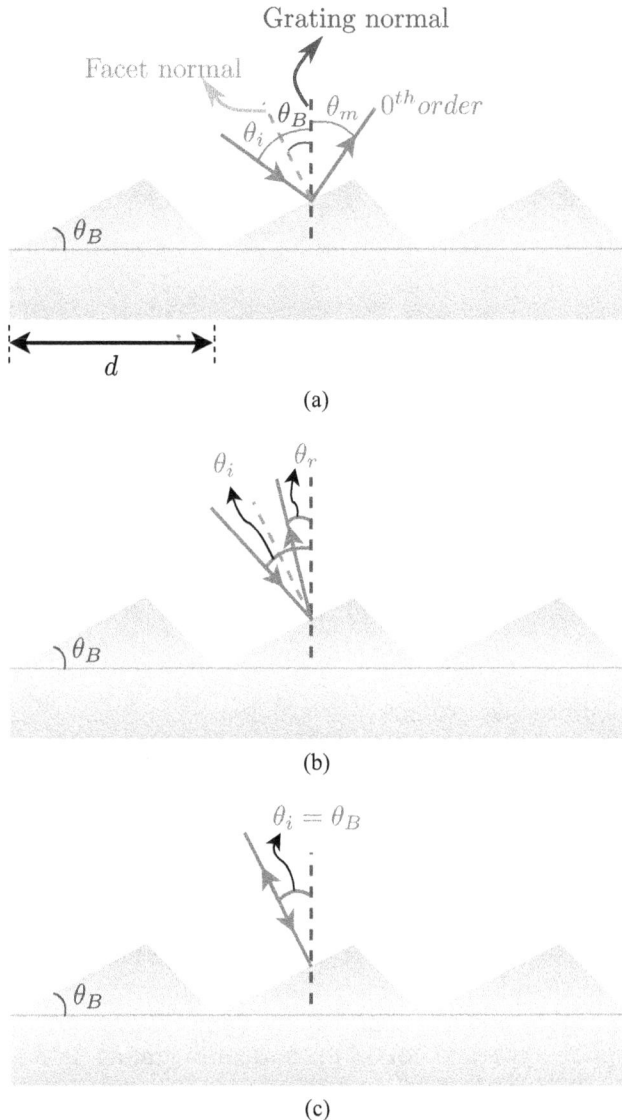

Figure 7.15. (a) Blazed grating, (b) specular reflection from facets of blazed grating and (c) Littrow incidence.

7.4 Designing diffractive optical elements

The diffraction grating discussed in earlier sections is an example of a DOE. It is in fact the first optical element that was meant to work on the principle of diffraction [7]. At the time of its early development, a lot of what is now considered fundamental physics was not yet understood, and the grating was studied more by a trial-and-error approach. Today, we understand the physics and have access to tools that help us design diffractive optics for specific applications. We know that phase affects an incident light beam. Any element, be it a refractive or diffractive one, has a spatially varying phase across it, which in turn manipulates the phase of this incident beam, changing its behaviour (either on transmission or after reflection). We can then think of an optical element as nothing other than the specific phase distribution required to carry out a particular job. For example, the plano-convex shape of a glass block focuses light. Or the shape of a concave field lens that collects light that would otherwise escape a system. The first question to ask in the design process is, 'How does one arrive at the required diffractive phase?' There are a number of ways of doing this. We describe three of the most common approaches.

7.5 Methods of generating a desired diffractive phase profile

7.5.1 Using a known refractive phase

The simplest way of generating a phase distribution is to start from a known refractive phase, such as the plano-convex lens shown in figure 7.16(a). A technique called the modulo 2π operation or the Fresnel technique is employed to convert that phase distribution to that shown in part (b) of the figure. As the name suggests, in a modulo 2π operation, one removes all integral multiples of 2π, and the element comprises only the remaining phase. Details of how exactly to do this are available in several references [8]. The element that remains is either a refractive or diffractive element based on the sizes of the smaller lateral features. If the sizes are $\gg \lambda$, the element is primarily refractive. Such large Fresnel lenses are commonly seen in lighthouses. When the feature sizes are of the order of or smaller than the wavelength, they are almost impossible to fabricate using standard grinding and milling tools, given the curved regions and abrupt height changes. This element is either converted to a multilevel or binary element. The former is not discussed in this

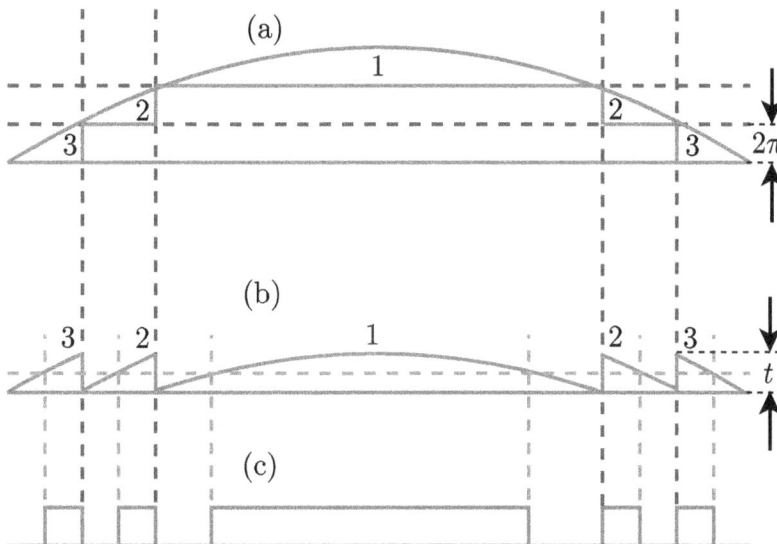

Figure 7.16. (a) Refractive lens, (b) analogue diffractive lens and (c) binary diffractive lens.

book. Suffice it to say it has a higher efficiency but requires more processing steps to fabricate than the binary element. For the rest of this chapter, we focus on binary elements with the purpose of introducing the basics of DOEs. Part (c) of the figure is a binary version of the element in part (b). It can be thought of as an approximation to the analogue element. The height of the element is chosen to ensure zeroth order cancellation. This element will still have the functionality of the original lens. However, several orders will be formed.

7.5.2 Using a known analytic expression

In the next chapter we will study complex beams in more detail. At this point, it is enough to note that each beam type has an expression that describes the changes its amplitude and phase undergo, as the beam propagates. We have observed this already when studying Gaussian beams. Another beam of interest is known as the Bessel beam and is described by the following equation:

$$E(r, z) = A\, J_0(k_r r)\mathrm{e}^{\mathrm{i}k_z z}, \tag{7.44}$$

where $r = \sqrt{x^2 + y^2}$, J_0 is the Bessel function of the zeroth order, and k_r and k_z are the radial and longitudinal wavevectors, respectively.

This expression can be used to create the phase of a diffractive element. However, the equation has terms that are a function of amplitude and phase. The binary elements that we have discussed so far consist of the transparent regions of a material of varying height. Such an element can affect only the phase in that vicinity, and not the amplitude. The way to incorporate amplitude variations is to use the technique of holography, as described in chapter 7 of [8]. Let us take equation (7.44) to be the object beam of the holographic process, as defined in equation (6.49). We can consider a tilted beam, similar to the one in equation (6.50) to be the reference beam. The two beams superpose and the interference intensity $I(x, y)$ is obtained. The phase of the corresponding holographic element is

$$\phi(x, y) = \exp[j\pi I(x, y)]. \tag{7.45}$$

The resulting phase could be loaded directly on a spatial light modulator (SLM) or binarised to enable easy fabrication using lithography. This technique can be used to convert any known analytic expression into a diffractive element. Illumination of the element with a tilted reference wave will generate the desired object beam, albeit in an off-axis direction.

7.5.3 Using an iterative technique

In certain cases, no analytic description of the desired beam exists. Two things are known, however. They are the input beam and the desired output intensity. Many computational techniques exist [9, 10] by which this information can be used to arrive at the phase function that will convert the former to the latter. The simplest algorithm is called the Gerberg–Saxton algorithm [11], known as the inverse Fourier transform algorithm (IFTA) method [12]. More details about how to implement this algorithm with actual code to do the same are available in [8].

7.6 Bridging the gap between theory and design tools

Whether calculating the Fresnel diffraction field using either equation (7.5) or (7.9) or the Fraunhofer diffraction field given by equation (7.11), usually a software tool such as Matlab® (a registered trade mark of The MathWorks, Inc.) will be used. The rest of this discussion talks about problems with the output calculated field that arise simply because of the way data are stored in such software. In Matlab, the field over a plane would be stored in a 2D matrix. This means that the field values are discretised, as they are calculated only at points. This idea is demonstrated in figure 7.17.

While both the source and image planes are discretised, it is important to note that separate variables are needed at each plane, as the spacing between points and the overall dimensions may not be the same. This is because as the beam propagates, its lateral extent changes. The Nyquist criterion must be satisfied at any plane over which the field is sampled. The challenge is to calculate the fields with adequate sampling, while capturing the entire diffracted field.

Both the sampling and lateral extent are dependent on the equation used to calculate the diffracted field in the output plane. We will use figure 7.18 to understand these differences, keeping in mind that Δx_{obs} is the lateral extent that is actually observed in the simulation in the output plane, whereas Δx_{act} is the true or actual transverse extent of the beam. The lateral extent of the source plane and the distance between samples in that plane are $\Delta \xi$ and $\delta \xi$, respectively. N_s and N_{out} denote the number of samples (in the x direction) at the source and output planes, respectively. The variable x is used in the image/output plane.

The differences that can arise when modelling the Fresnel diffraction field can broadly be classified as follows [13]; in each case, the assumption is that beam

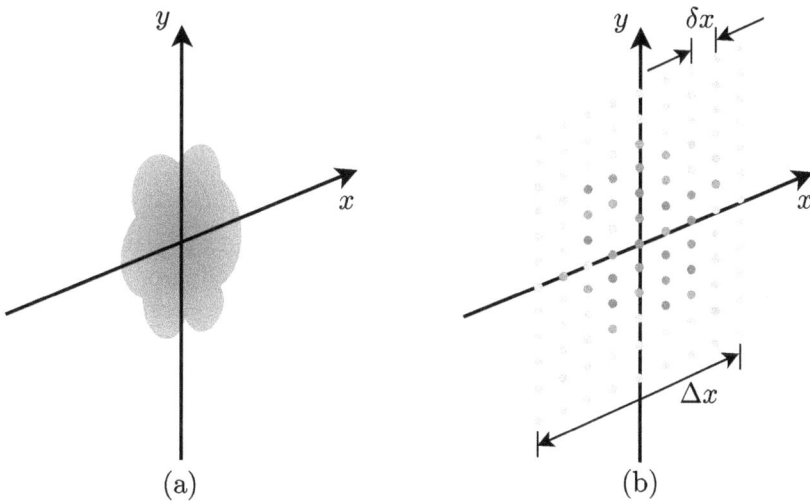

Figure 7.17. Field across plane of interest: (a) continuous values and (b) in the x direction, N discrete values with a sampling spacing of δx and overall extent of beam Δx.

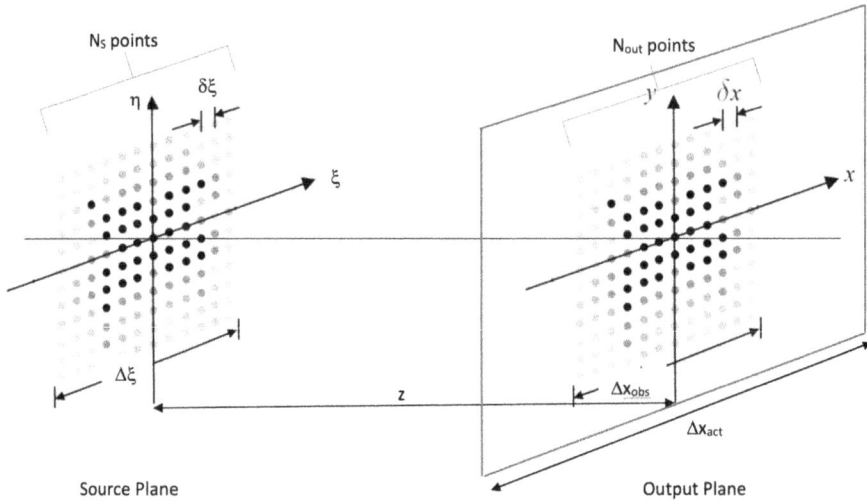

Figure 7.18. Figure showing a change in parameters at starting and final planes of interest. In the output plane, the actual diffraction field Δx_{act} might cover a region larger than what is observed Δx_{obs}. The actual field is shown as a red box in the figure.

propagates from one plane to another, and Fresnel equations are being used to calculate the field at the second (output) plane:

1. The number of samples remains constant, i.e. $N_{\text{out}} = N_s$ but the sample size increases with propagation, $\delta x > \delta \xi$. As $\Delta x_{\text{obs}} = \Delta x_{\text{act}}$, the entire output plane is probed but may be under-sampled. This is the case when using equation (7.5).

2. In the second case, both the number of samples and the sampling size remain constant, $N_{\text{out}} = N_s$ and $\delta x = \delta \xi$. This means the sampling rate is constant irrespective of the propagation distance. Since $\Delta x_{\text{obs}} = \delta x \times N_{\text{out}} < \Delta x_{\text{act}}$ only a portion of the actual plane is observed, albeit with good sampling. As z increases, less of the output plane is probed. This is the case when equation (7.9) is used to calculate the field.

3. In the first case, the output plane was under-sampled as the number of samples remained constant. Therefore, in this third case, the number is increased to ensure that the output plane is properly sampled. In programming, this is achieved by zero-padding the starting plane to match the requirement in the output plane, as shown in figure 7.19. The useful information in the source plane (indicated by the orange square in the centre) is contained within a smaller area of the plane, sampled with N_s points (in the x direction) but the matrix itself has N_{out} points, such that $N_{\text{out}} > N_s$ in the x direction. The remaining points (in green) are simply zeroes that carry no useful information. They can be thought of as place holders for the lateral information that will become available in the output plane (shown in pink), the entire region of which contains useful information.

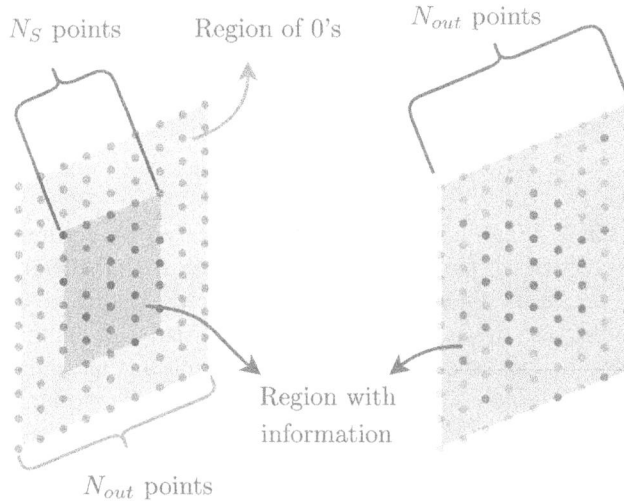

Figure 7.19. Figure showing how zero-padding the source plane provides more information in the output plane.

Table 7.1. Comparison of parameters when using different Fresnel diffraction equations for calculating the field at an output plane of double the lateral extent compared to the starting or source plane, i.e. $\Delta x = 2 \times \Delta \xi$. δx and Δx have units of length.

Equation	δx	N_{out}	Δx_{obs}	Δx_{act}
(7.5)	2	100	200	200
(7.9)	1	100	100	200
(7.5) and zero-padding	1	200	200	200

To illustrate the differences more clearly, an example is taken where at the source plane, the values of the parameters are assumed to be $N_s = 100$, $\delta \xi = 1$ and $\Delta \xi = 100$, with the latter two parameters having units of length. Depending on the equations used to calculate the diffracted field, values of the parameters of interest at the output plane are given in table 7.1.

From table 7.1, one can see that in the first case the entire output plane is probed ($\Delta x_{obs} = \Delta x_{act}$) but with lower sampling ($\delta x = 2 >$ the sampling at the input plane). On the other hand, the sampling is good in the second case but only a part of the output plane is probed ($\Delta x_{obs} < \Delta x_{act}$). The final case is the best one where the entire output plane is sampled with good sampling.

One more point to be noted is that as the diffracted field is calculated using Fourier transform equations, such as equations (7.5), (7.9) and (7.11), the coordinates in the source and output planes represent space and spatial frequency, respectively. If the length of one side of the plane is taken as L and the number of sampling points as N, then the sampled space in both these planes can be programmed as

- $\delta\xi = L/N$: source plane sample interval.
- $\xi \rightarrow [-\frac{L}{2}: \delta\xi: \frac{L}{2} - \delta\xi]$: sampling space along one dimension.
- $fx \rightarrow [-\frac{1}{2\delta\xi}: \frac{1}{L}: \frac{1}{2\delta\xi} - \frac{1}{L}]$: corresponding frequency coordinates.

This section has highlighted some of the issues faced by commonly used beam propagation techniques. In particular, it should be clear that the angular spectrum method is not the best choice when the beam needs to propagate over a long distance. This is because the increasing (lateral) field size with distance means that a smaller part of the output field is being imaged at each further plane. To overcome this problem, zero-padding is carried out but this reduces the amount of information available in the input plane. An improved angular spectrum method [14] can be employed to overcome these issues.

7.7 Problems

Use Matlab or a similar tool to simulate the following diffraction fields. Helpful information and even some code are available in [8].

1. Create a square aperture of side 25 pixels at the centre of a 512×512 matrix. View the far-field diffraction pattern. How does it change if the aperture size changes to 50 pixels?
2. Create two square apertures, each of side 20 pixels, within the 512 pixel matrix. Let the spacing between them be 5 pixels. Simulate the diffraction pattern of this configuration and notice the difference in the diffraction pattern.
3. Vary the distance between the slits and add more slits. In each case, how does the far-field pattern change?
4. Simulate a 1D grating of 50% duty cycle and observe the far-field pattern. Change the duty cycle to 10%, 30% and 90%. Observe the far-field in each case. What can you say about the efficiency of the first order in the 10% and 90% cases?
5. Assuming a wavelength of 633 nm is incident on a 1D grating (side 1000 pixels and period 20 pixels), calculate the Fresnel diffraction at a distance of 1.6 mm away using both equations (7.5) and (7.9). What is the sampling in each case? What happens if the field is observed only 1 mm away?
6. A student is curious about how much information is stored on a CD compared to a DVD. She carries out an experiment to measure the difference in spacing of the data tracks in both cases. The set-up is as shown in figure 7.20. She uses a laser pointer of 650 nm wavelength. The light falls normally on the disk. She sets the distance z from the disk to the observation screen to be 10 cm. She then measures the distance between the first and zero orders on the screen to be 4.5 cm and 18.5 cm in the case of the CD and DVD, respectively. What is the track spacing on each storage medium? Which one is smaller? DVD systems use a laser of wavelength 650 nm but

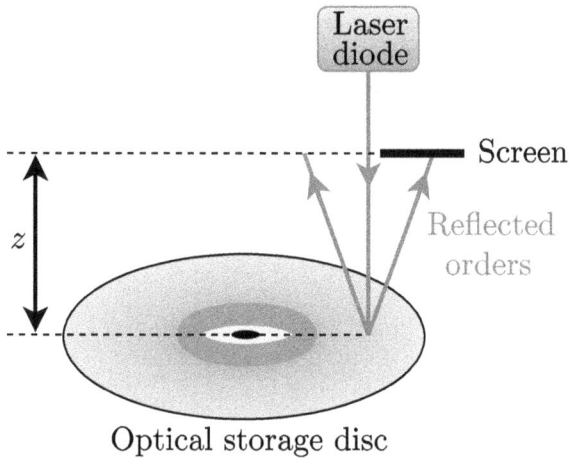

Figure 7.20. Experimental set-up to measure the distance between tracks on optical storage media.

CD systems use a larger wavelength, namely 780 nm. What difference does this make? (Try this experiment at home! You only need a laser pointer and a CD or DVD. Remember not to look at the laser light directly and be careful of all the diffracted orders and reflected light.)

References

[1] Huygens C 1912 *Traité de la Lumière* (Leyden: van der Aa) https://archive.org/details/huyghens-traite-de-la-lumiere-gauthier-villars-1690-english-trans/page/n1/mode/2up
[2] Miller D A B 1991 Huygens's wave propagation principle corrected *Opt. Lett.* **16** 1370–2
[3] Goodman J W 1996 *Introduction to Fourier Optics* (New York: McGraw-Hill)
[4] Hell S W 2014 Nobel Lecture http://www.nobelprize.org/prizes/chemistry/2014/hell/lecture/ (Accessed: 17 June 2023)
[5] Oppenheim A V, Willsky A S and Nawab S H 1997 *Signals and Systems* Prentice-Hall Signal Processing Series (Englewood Cliffs, NJ: Prentice-Hall)
[6] Palmer C A and Loewen E G 2005 *Diffraction Grating Handbook* (Irvine, CA: Newport)
[7] Hazra L 1999 Diffractive optical elements: past, present, and future *Proc. SPIE* **3729** 198–211
[8] Vijayakumar A and Bhattacharya S 2017 *Design and Fabrication of Diffractive Optical Elements with MATLAB* (Bellingham, WA: SPIE)
[9] Yoshikawa N and Yatagai T 1994 Phase optimization of a kinoform by simulated annealing *Appl. Opt.* **33** 863–8
[10] Seldowitz M A, Allebach J P and Sweeney D W 1987 Synthesis of digital holograms by direct binary search *Appl. Opt.* **26** 2788–98
[11] Gerchberg R W 1972 A practical algorithm for the determination of phase from image and diffraction plane pictures *Optik* **35** 237–46
[12] Fienup J R 1978 Reconstruction of an object from the modulus of its Fourier transform *Opt. Lett.* **3** 27–9

[13] Mas D, Garcia J, Ferreira C, Bernardo L M and Marinho F 1999 Fast algorithms for free-space diffraction patterns calculation *Opt. Commun.* **164** 233–45
[14] Matsushima K and Shimobaba T 2009 Band-limited angular spectrum method for numerical simulation of free-space propagation in far and near fields *Opt. Exp.* **17** 19662–73

IOP Publishing

Introduction to Ray, Wave, and Beam Optics with Applications

Shanti Bhattacharya

Chapter 8

Introduction to complex light

The simplest solutions to the wave equation are usually written as electric fields, whose general form is given by equations (1.1) or (6.1). They provide information about the frequency (or wavelength), velocity and polarisation of the light wave. Based on these equations, one might argue that light has always been complex, as the equations that represent it contain both amplitude and phase information. However, the term complex light has come to mean something more than the argument of the exponential of the electric field. In its current avatar, complex (or structured) light describes beams with unusual and, sometimes, inhomogeneous properties. These properties could include special intensity distributions, cross sections with radially varying polarisation, beams with orbital angular momentum or even beams that appear to accelerate as they propagate. There are many different types of complex light beams including Bessel [1], Laguerre–Gaussian [2], Ince–Gaussian [3], Mathieu [4] beams, etc. All beams can be arrived at by solving the wave equation under a specific set of conditions. The theory behind these beams is beyond the scope of this textbook. Instead, a more intuitive approach to how they exist is presented. We take three very distinct examples of such beams, explore their properties, see how they can be generated using diffractive optics, and give a brief introduction to the applications they are transforming or making possible.

8.1 Bessel beams

The BB [5] is one of the solutions to the Helmholtz wave equation given in equation (5.3). They have interesting properties that make them useful for a variety of applications. The Bessel beam is described by

$$E(r, z) = AJ_0(k_r r)e^{ik_z z},\tag{8.1}$$

as already mentioned in the previous chapter. The radial and longitudinal wave vectors are related through the expression $k = \sqrt{k_r^2 + k_z^2}$.

doi:10.1088/978-0-7503-5497-4ch8

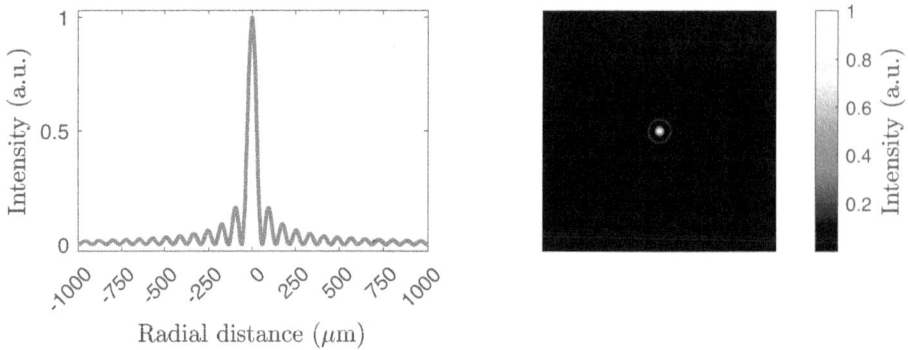

Figure 8.1. Bessel beam intensity. The figure on the left is a line scan (through the centre of the image) of the cross section of the beam shown on the right. Picture credit: JG.

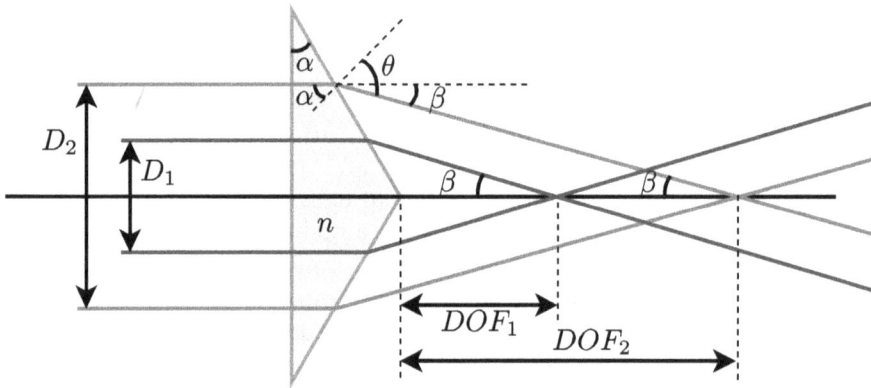

Figure 8.2. Effect of the size of the beam on the depth of focus of a Bessel beam.

A single Bessel beam can be considered to consist of a set of plane waves propagating on a cone [1, 6, 7]. A cross section of the beam in the near-field consists of a narrow high-intensity peak, surrounded by rings as seen in figure 8.1.

8.1.1 Propagation-invariance

We have seen how diffraction causes a travelling wave to spread. The ideal Bessel beam has infinite energy and can propagate an infinite distance without diverging. It is this property that is referred to as propagation-invariance. Practical Bessel beams have a reduced divergence over a certain distance known as the depth of focus, which is much less than infinity. We will see that the DOF of a Bessel beam will be longer than the corresponding Rayleigh range of a Gaussian beam of similar size. To understand why this is, we look at how these beams are generated. An element called an axicon or a conical lens is used to create Bessel beams. A cross section of an axicon, with opening angle α, is drawn in figure 8.2. The figure also has beams of two different diameters $D_1 = 2R_1$ and $D_2 = 2R_2$ incident on the axicon. The material of the axicon has refractive index n.

From the figure, it appears that the DOF depends on both the size of the axicon (assuming the beam is of the same size) and the angle β. The latter in turn depends on the the angle of refraction of the light θ after having travelled through the axicon. In general, for an axicon of size $D = 2R$, the following is true:

$$\tan \beta \approx \frac{R}{\text{DOF}} \qquad (8.2)$$

or

$$\text{DOF} = \frac{R}{\tan \beta} = \frac{R}{\tan(\theta - \alpha)}. \qquad (8.3)$$

Snell's law tells us that

$$n \sin \alpha = \sin \theta,$$

assuming that the axicon is surrounded by air. Substituting this into the equation for the DOF yields

$$\text{DOF} = \frac{R}{\tan(\sin^{-1}(n \sin \alpha) - \alpha)}. \qquad (8.4)$$

Equation (8.4) validates the belief that the DOF is a function of both the radius of the incident beam and the cone angle of the axicon. If R were infinite, then so would the DOF. Of course, this would imply the beam had infinite energy. The graph plotted in figure 8.3 enhances our understanding of the effect of the cone angle has on the DOF.

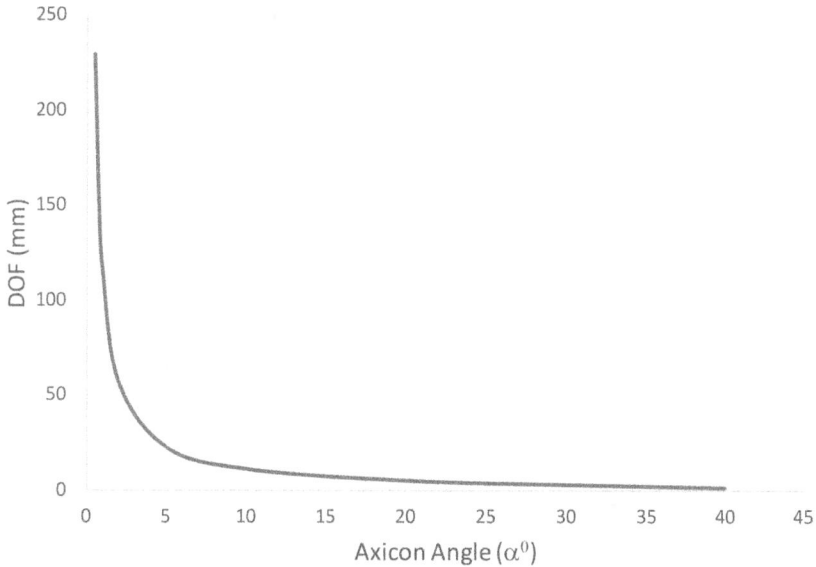

Figure 8.3. Variation of the depth of focus with axicon angle obtained by plotting equation (8.4). The diameter of the beam incident on the axicon was set at 2 mm and the refractive index at 1.5. Data generated by SB.

8-3

There are a couple points of interest to note from this graph:
- If the axicon angle could be made extremely small, even a finite-sized beam would have an extremely large DOF.
- Beyond $\alpha = 40°$, the DOF is negligible but it is already quite small even for $\alpha > 15°$.

The main aim of generating a BB is to extend the DOF of the beam. Therefore, axicons typically have very small cone angles. Because of this, refractive axicons can be quite challenging to fabricate.

Point to ponder: In figure 8.3, at large axicon angles, the DOF $\rightarrow 0$. What happens to the light if the axicon angle increases even further, i.e. $\alpha \gg 40°$?

Given that α is very small both $\tan \alpha$ and $\sin \alpha \approx \alpha$, and equation (8.4) can be simplified to

$$\text{DOF} = \frac{R}{\alpha(n-1)}. \tag{8.5}$$

To really appreciate a Bessel beam, we compare it with a Gaussian beam of identical wavelength, whose diameter matches the width of the central peak of the BB. The radius of the central core of a Bessel beam [6] is obtained from the location of the first zero of the Bessel function is and given by

$$r_0 = \frac{2.405}{k_r}. \tag{8.6}$$

The difference between the Gaussian beam's Rayleigh range and the BB's DOF can be observed in figure 8.4.

We see that the BB can travel further with less divergence than a Gaussian beam of similar size. It is this property that makes the BB so useful in various applications.

8.1.2 Self-healing

Another interesting property of Bessel beams is their ability to recover after encountering an obstacle in their path, a property known as self-healing or self-reconstruction [8]. The first image in figure 8.5 shows the cross section of a Bessel beam, with the central part blocked at plane z_1. The subsequent images are cross sections of the same beam at different points along the optical axis beyond the disturbance.

It can be seen that within a short distance, the beam has reconstructed itself. The self-healing distance will depend on the size of the obstacle and the wave vectors of the beam [6]. Why is the Bessel beam less affected by an obstacle in its path and how

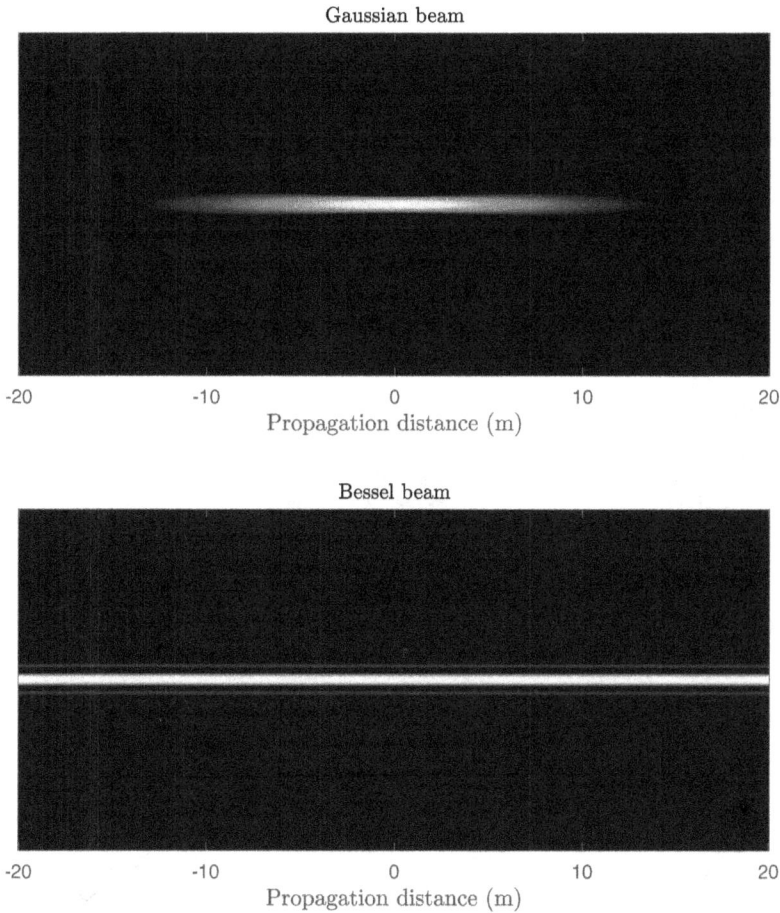

Figure 8.4. Comparison between GB Rayleigh range and BB depth of focus. Picture credit: JG.

does it recover? It is helpful to compare the BB to a Gaussian beam to answer this. In section 5.2.3, we saw that a GB contains 86% of power within a circle defined by the GB radius. On the other hand, the power in a finite-sized Bessel beam is equally divided between the central core and surrounding rings. The more rings a beam has, the larger its DOF but the less the power in the core. This means that the fraction of the total power in the central core is much less than that of a Gaussian beam. Therefore, if an obstacle blocks it, only a small part of the power is lost. As the beam travels after the obstacle, energy from the rings feeds back into the central part of the beam, reconstructing the Bessel beam core. This property makes BBs very useful in imaging samples such as tissue which is highly scattering [9].

8.1.3 Intensity of a Bessel beam along the direction of propagation

Bessel beams are probably the form of complex light that most people are familiar with. There are numerous publications about them and their applications. Their

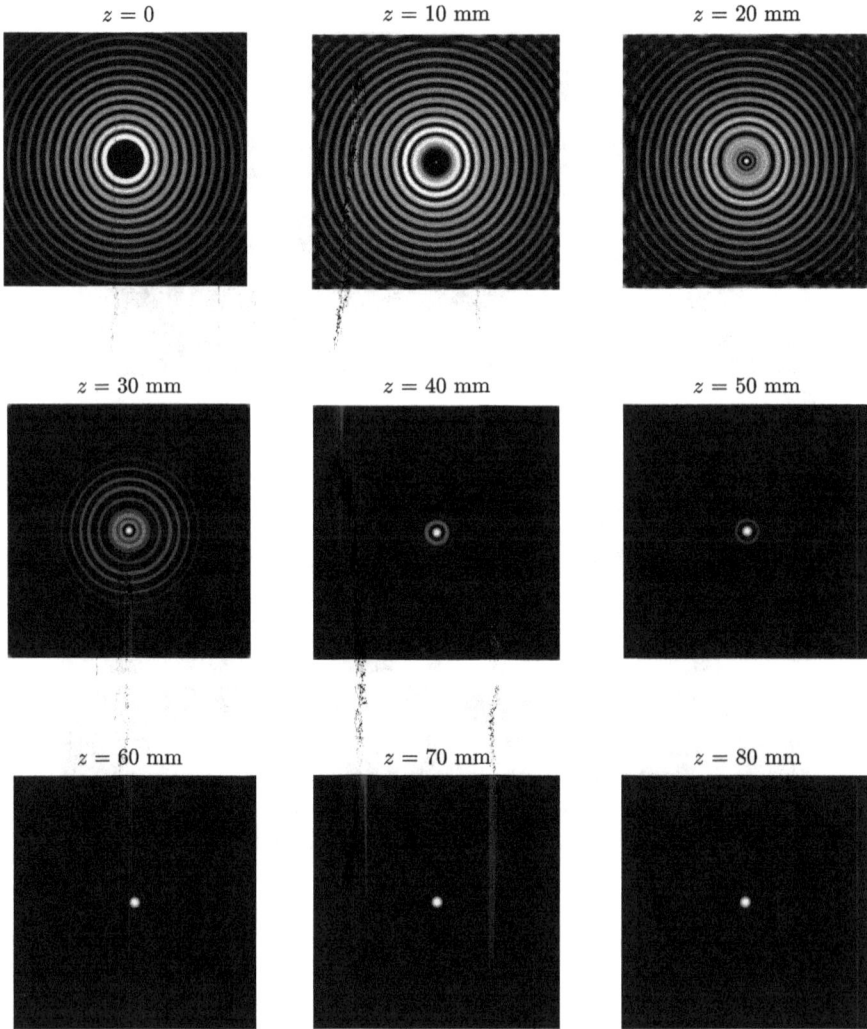

Figure 8.5. Self-healing of a Bessel beam. For this simulation, the diameter of the axicon and core radius of the Bessel beam were 2 mm and 60 μm, respectively. The diameter of the block was 300 μm, and the wavelength of the light was assumed to be 633 nm. The block is placed at $z = 0$, and the beam allowed to propagate beyond that. Picture credit: JG.

propagation-invariance and self-healing properties are well known. What is often not discussed or even mentioned is the fact that the intensity of the central peak of the beam, even within the DOF, is not constant. Careful observation of figure 8.2 should explain why this is so. From this figure, we can see that the rays incident on each annular zone of the axicon contribute to the intensity at particular position on the axis. This is explained nicely in figure 1 of [10].

Since the intensity variation along the axis after the axicon is a function of the transverse intensity variation incident on the axicon, changing the latter should

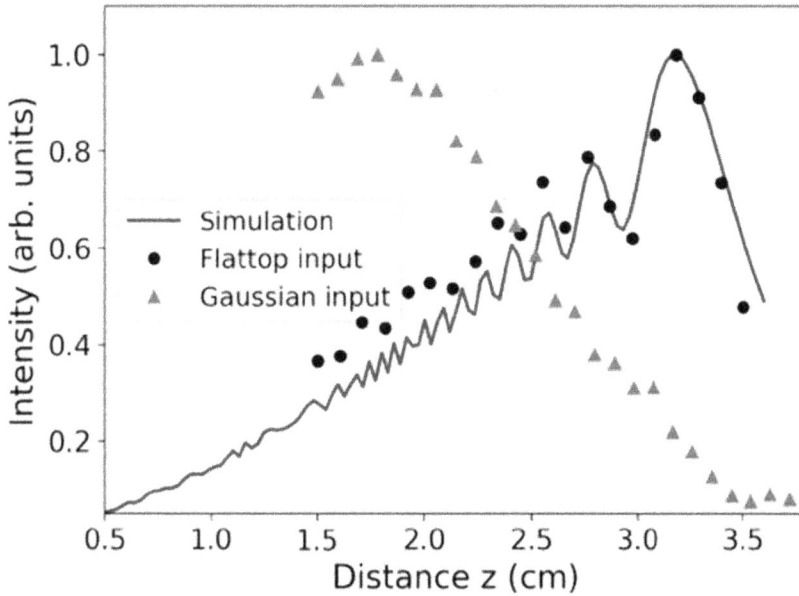

Figure 8.6. Variation of axial intensity along the optical path after an axicon for different incident profiles. The solid line shows the simulated and dotted circles show the experimentally measured axial intensity when a flattop beam falls on a diffractive axicon. The green triangles show the experimental results for an incident Gaussian beam. Reproduced with permission from [10]. Copyright 2018 IOP Publishing.

control the former. The same paper shows that one can even increase the peak intensity of the Bessel beam as it travels away from the axicon by controlling the incident intensity profile. Figure 8.6 shows how the peak intensity varies along the axis when the intensity profile of the incident beam has been altered. The intensity of a Gaussian beam was transformed (using a DOE) to have a flat-top intensity. This was incident on a diffractive axicon resulting in a growing peak intensity, as the BB travels away from the axicon. The ripple in the intensity is caused by diffraction due to the sharp point at the centre of the axicon. Diffractive elements were used to both control the intensity profile of the incident light, as well as create the axicon. For comparison, the axial variation in intensity when a Gaussian beam is incident on the same axicon is also shown in the figure.

8.1.4 Far-field pattern of a Bessel beam

What would the far-field distribution of a Bessel beam look like? To answer that, we recollect that the FT process provides information on the spatial frequencies present in the signal, which in this case is the Bessel beam. Every plane wave at an angle to the optical axis represents a specific spatial frequency. The FT of a single tilted plane wave would result in a point, at a distance away from the axis, and whose location would be determined by the tilt angle. Since a Bessel beam consists of an infinite set of plane waves propagating along a cone, its far-field pattern, which is the FT of equation (7.44), consists of an annular ring. It is interesting to note that the first element used to generate a Bessel beam was a transparent annular ring [6]. Apart from the light that travels through the ring, the rest is cut off by the amplitude element. Therefore, this is not a very efficient method of generating a Bessel beam, which is why refractive, diffractive or meta-optical phase axicons are preferred means by which to generate a BB.

8.1.5 Laguerre–Gaussian beams

The equation for a Laguerre–Gaussian (LG) beam [11] is

$$\mathrm{LG}_{lp}(\rho, \phi, z) = C_{lp}^{\mathrm{LG}} \exp\left(\frac{-\rho^2}{w^2}\right)\left(\frac{\sqrt{2}\,\rho}{w}\right)^{|l|} \mathrm{LG}_p^{|l|}\left(\frac{2\rho^2}{w^2}\right)\exp\left(jl\phi + \frac{jk\rho^2}{2R} - j\psi_{\mathrm{G}}\right), \quad (8.7)$$

where C_{lp}^{LG} is a normalisation constant [12], R and w are the radius of curvature and beam radius. They are defined exactly as they are for a Gaussian beam, namely through equations (5.10) and (5.11). LG is the Laguerre polynomial [13], and l and p are the azimuthal and radial modal numbers. $l = 0$ represents the well-known Gaussian beam. Therefore, it should not be surprising that LG beams have a Gouy phase associated with them, given by

$$\psi_{\mathrm{G}} = (2p + |l| + 1)\tan^{-1}(z/z_0). \quad (8.8)$$

It is interesting to compare this with the Gouy phase of a Gaussian beam given in equation (5.13). The main difference of the LG beam from a GB is the $\exp(jl\phi)$ term in the phase, which introduces the azimuthal phase. The implication of this is that a small dielectric object placed along the path of the beam will experience a torque or orbital angular momentum, which is why these beams are also sometimes referred to as twisted light. Orbital angular momentum (OAM) should not be confused with spin angular momentum (SAM), which relates to the polarisation of a light wave. The difference in the effect of each of these on a photon is shown in figure 8.7.

A striking difference of this beam from most other beams are that it has one continuous wavefront that sweeps out a helical surface as it travels, as seen in

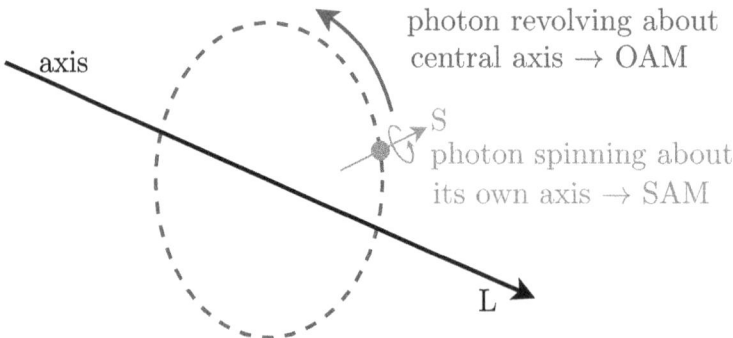

Figure 8.7. Difference between OAM and SAM.

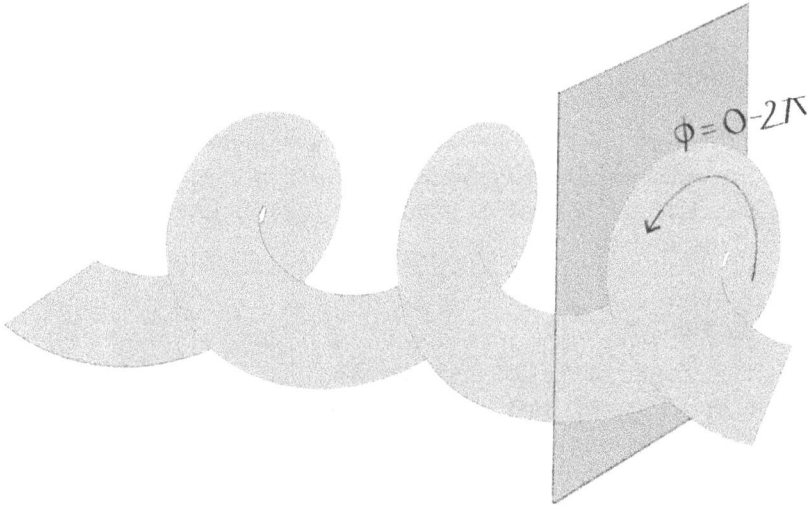

Figure 8.8. The wavefronts of a beam with orbital angular momentum are shown. The grey cross section of this image is not a wavefront but results in the familiar picture of an OAM beam with the 0–2π phase variation. Picture credit: SumiB.

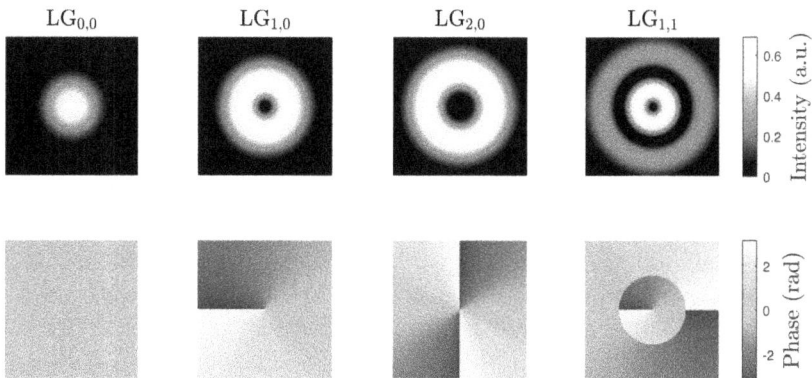

Figure 8.9. Intensity and phase distributions for different combinations of l and p of a Laguerre–Gaussian beam. Picture credit: JG.

figure 8.8. This is very different from a plane, spherical or Gaussian wave that has distinct independent wavefronts along the direction of propagation.

If the phase was measured in a planar section across the beam, it would be found to be varying from 0 to $m2\pi$ for a beam with $l = m$, where m is an integer. Figure 8.9 displays the intensity and phase distributions for a few different combinations of l and p.

Point to ponder: Why do the intensity distributions of the OAM beams (with $l > 0$) have a null intensity at the centre?

Looking at the phase distribution should give a us a clue to answering this. Since the phase is constant along any one radial direction, it is undefined at the centre. Such points in a beam are called singularities. The intensity null is the reason that OAM and LG beams are sometimes referred to as donut beams. The name vortex beam, on the other hand, arises from their spiral phase. OAM beams are not propagation-invariant. They are of great interest in applications such as optical trapping and manipulation [14, 15].

8.2 Airy beams

In the very first chapter of this book, we discussed one of the fundamental postulates of geometric optics, namely that light travels in straight lines. The Airy beam is a very interesting solution to the wave equation which seems to violate this principle. Airy beams are found to accelerate or follow a curved path, as they propagate. This can be seen in figure 8.10(a). A cross section at a particular value of z yields the asymmetric amplitude and intensity pattern as shown in figure 8.10(b).

It turns out that the centroid of the transverse intensity always moves in a straight line. However, the features of the intensity profile of an Airy beam follow a curved trajectory. Figures 8.11(a) and (b) clearly demonstrate this behaviour. One can think of the overall momentum of the beam being conserved, as the maximum intensity moves in one direction while the intensity in the (transverse) tail of the beam spreads out to the opposite side.

It has even been shown that the peaks of Airy beams behave like projectiles under the influence of gravity [16]. The idea of such self-accelerating waves was first proposed in 1979, not in the field of optics but in quantum mechanics by Berry and Balazs [17]. They found that such waves were solutions of the potential-free Schrödinger equation. The interesting thing was that these solutions maintained their shape on propagation [18]. Optical Airy beams were demonstrated several decades later [19, 20]. These researchers looked specifically at the paraxial wave

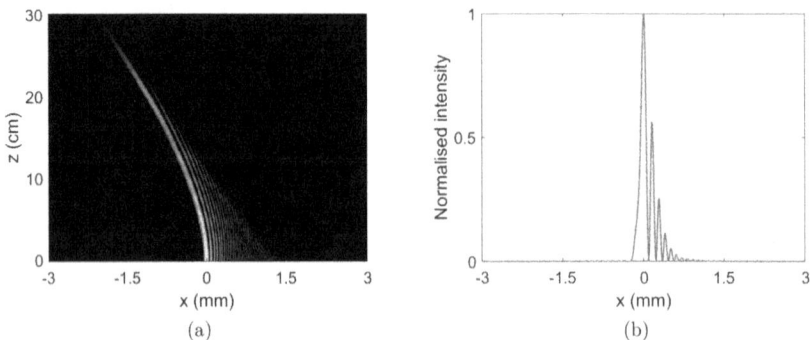

Figure 8.10. (a) A view of a 1D Airy beam as it propagates. (b) A cross section (of the intensity) of the same beam, at one particular z position. Picture credit: JG.

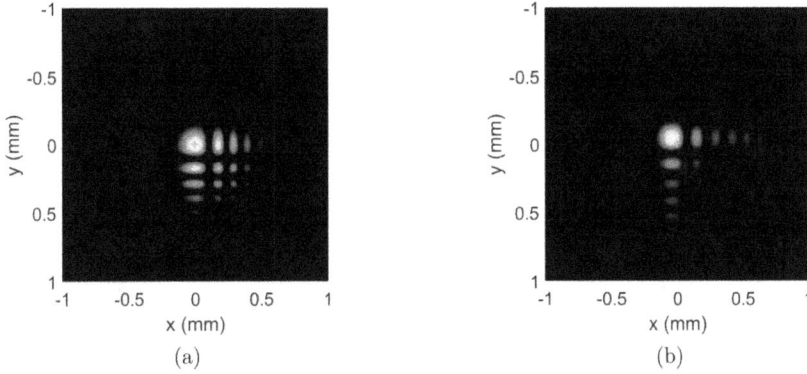

Figure 8.11. A cross section of a 2D Airy beam at (a) $z = 0$. (b) And the same beam after it has traversed some distance. Picture credit: JG.

equation, equation (5.4), in one dimension. Therefore, setting $\frac{\partial^2}{\partial^2 y} = 0$ allows this to be rewritten as

$$\frac{1}{2k}\left[\frac{\partial^2}{\partial^2 x}\right]E + j\frac{\partial E}{\partial z} = 0. \tag{8.9}$$

The Airy solution can be arrived at in the case when the following substitutions are made [19]:
- First, the field being solved for is represented by the function $\phi(s, \xi)$ instead of $E(x, z)$,
- ϕ is a function of a dimensionless transverse coordinate s and a normalised propagation variable ξ, i.e. $s = x/x_0$ and $\xi = z/(kx_0^2)$ and
- x_0 is an arbitrary transverse scale factor.

For finite-sized Airy beams, $\phi(s, \xi) = \text{Ai}(s)\exp(as)$, where Ai represents an Airy function [21].

Airy beams fall under the class of non-diffracting, self-healing beams and have found applications in optical imaging [22].

8.2.1 Far-field pattern of an Airy beam

The Fourier transform of an Airy beam results in a Gaussian beam with an extra phase term that contains a cubic phase [18], i.e.

$$\text{phase} \propto \exp\left[\frac{2\pi}{\lambda}\left(x^3 + y^3\right)\right]. \tag{8.10}$$

This phase distribution is displayed in figure 8.12.

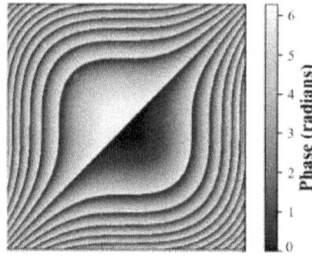

Figure 8.12. Cubic phase profile in the far-field of an Airy beam. Reproduced with permission from [23].

8.3 Generation of complex light

Complex light generation involves the conversion of a beam (e.g. a flat-top intensity collimated wave or a Gaussian beam) into one with some desired properties. Some complex light can be generated with refractive optics, e.g. an axicon can generate a Bessel beam. However, other complex light requires elements that cannot be fabricated using conventional grinding or polishing tools. In such cases, diffractive or meta-optical elements are used to generate the light. It should be noted that a spatial light modulator (SLM) can always be used as well. These devices are tunable and the phase they carry can be changed, as and when required. In section 7.4, we studied different ways that diffractive elements could be designed. All of these are valid means by which to create complex-light-generating phase elements. In the following sections, we look at diffractive elements that generate the specific examples of complex light just discussed.

8.3.1 Bessel beams from diffractive axicons

The phase of a diffractive axicon can be obtained from knowledge of the phase profile of a refractive axicon. Using the modulo 2π operation would create a binary element, a cross section of which is shown in figure 8.13. An interesting point about a binary diffractive axicon is that one can obtain an efficiency of $\approx 80\%$, as both the ± 1 orders contribute to the Bessel beam formation.

8.3.2 OAM beams from fork gratings

In theory, one can fabricate refractive spiral phase plates to generate OAM beams, as seen in figure 8.8, but machining such an element is extremely challenging. Instead, a holographic technique is used. The analytic expression of the desired OAM beam is interfered with a tilted plane wave and the resulting intensity creates a phase pattern as given in equation (7.45). In the case of an $l = 1$, the OAM—tilted plane wave interference pattern obtained resembles a grating with a fork at the location of the singularity. The phase element of course has this same pattern, as seen in figure 8.14.

Since this is a holographic technique, the desired OAM beam will be generated off-axis. More recently, researchers have created meta-optical elements that generate focused OAM beams [24], which will be generated on-axis.

Figure 8.13. Cross section of a binary diffractive axicon. The grey and green regions represent regions of 0 and π phase, respectively. Reproduced with permission from [23].

Figure 8.14. Cross section of a binary diffractive fork grating. The black and white regions represent regions of 0 and π phase, respectively.

8.3.3 Airy beams from cubic phase plates

Airy beams can be generated from a binary diffractive pattern, whose phase distribution can be obtained by binarising the phase given in figure 8.12.

8.4 Applications

We have almost come to the end of this book! We can now start looking at applications that use all the ideas that have been explored and discussed so far. In the remaining part of this chapter, we explore a few applications that have benefited greatly from or have only been made possible by the use of complex light.

8.4.1 Optical trapping

Optical trapping, also known as optical tweezers, is a technique that uses a tightly focused laser light to trap and manipulate microscopic objects, such as small particles or cells. It was first demonstrated by Ashkin [25], for which he was awarded the Nobel Prize in Physics in 2018. Optical traps are of interest for a variety of reasons. For example, cells can be held by an optical tweezer and the properties of their DNA and RNA can be studied. The folding of proteins can be studied, as changes in this process are a good diagnostic tool to identify various diseases, and the forces of molecular motors can be studied and quantified, helping with the understanding of movement in living organisms.

The basic idea behind optical trapping is the fact that light carries momentum. This means that a particle interacting with a beam of light will experience a force. Refraction and reflection change the direction of light, causing a change in the momentum of the light. Newton's third law of motion tells us that the object must undergo an equal and opposite momentum change. The large intensity variation across a tightly focused beam creates a gradient force on a small transparent particle close to the focus. The intensity gradient creates a region of high intensity to which the object is attracted. This volume in space is referred to as the optical tweezer or trap. The principle is shown in figure 8.15. The particle being trapped is considered to be a transparent dielectric.

In the figure, the resulting net momentum is also drawn. It can be seen to be pushing the particle towards the focus point. If the particle were in the centre, then the forces on either side would cancel and the particle would remain in the centre. If it moves to either side, it gets pushed back to the centre. The trapping mechanism relies on the fact that the intensity of the focused laser beam decreases rapidly as one

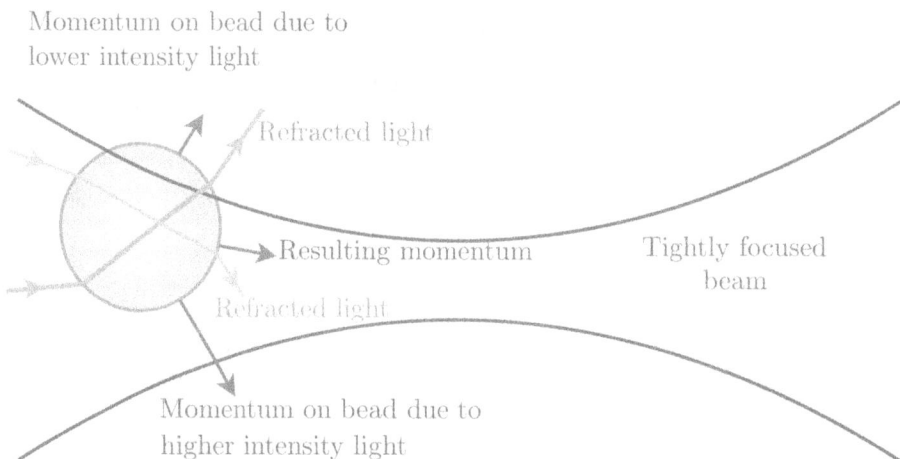

Figure 8.15. A particle is shown towards one side of a tightly focused beam. Light is incident all over the particle but two particular sets of rays are shown. The pink lines indicate a region of lower intensity, while the orange shows the higher intensity region. The appropriately coloured refracted light and resulting momentum directions (shown in green) are drawn in both cases.

moves away from the centre of the beam. As a result, particles tend to be trapped, at the highest intensity point, in the two-dimensional focal plane.

While optical traps can work with Gaussian beams, those that use focused Bessel beams, instead of tightly focused Gaussian beams, provide more flexibility for manipulation of particles in a three-dimensional volume corresponding to the DOF of the BB [26].

8.4.2 STED

In this example, we discover how an OAM beam generated by a diffractive element helped beat the diffraction limit. The resolution of any imaging system is limited by the Airy disk, whose radius is given by equation (2.16). When two Airy patterns are no longer distinguishable from each other, the system limit has been reached, and imaging of smaller features is not possible. To do so would require the optical system to create a smaller focused spot, which is exactly what a stimulated emission depletion (STED) microscope does. At the heart of an STED system is a laser scanning confocal fluorescent microscope laser scanning confocal microscopy (LSCM). Let us look at this system first before trying to understand the STED microscope. The LSCM contains the word confocal, as only light from the focus spot is allowed to reach the detector. This is made possible by using a pinhole to block any light from out-of-focus regions. This optical imaging system uses an objective to focus the laser light of wavelength λ_e onto a small spot (the size of the systems Airy disk) onto the sample. This laser can be considered to be the excitation signal. The sample must either have natural fluorophores within it or have had them added to enhance imaging. All the fluorophores within the focused spot are excited by λ_e and emit fluorescence at wavelength λ_f. The system collects the reflected light and with the help of a chromatic filter is able to separate the (desired) fluorescent light from the excitation wavelength. The collected fluorescent light provides information about one point of the sample. To create an image of the entire sample, either the sample or the focused spot is moved, hence the term *scanning* confocal microscopy.

The working principle of an STED microscope is shown in figure 8.16. A smaller focus spot is created by sending in two coaxially travelling beams of different wavelengths in quick succession.

Initially, the excitation laser beam (of diameter D_E) is sent in but this is followed by an OAM beam (of wavelength λ_{STED}, where $\lambda_{STED} > \lambda_e$) with a donut-shaped intensity profile. The fluorophores that had been excited by λ_e and are now exposed to the donut-shaped STED beam undergo stimulated emission. The outer diameter of the donut-shaped beam is D_{STED}, whereas the inner hole has a diameter D_D. Fluorescence now takes place only from fluorophores that lie within the hole of the donut. As can be seen in figure 8.16, this happens over a much smaller spot than the originally focused beam, i.e. $D_D < D_E$. Since the resolution is improved beyond the diffraction limit, this technique is called a super-resolution method.

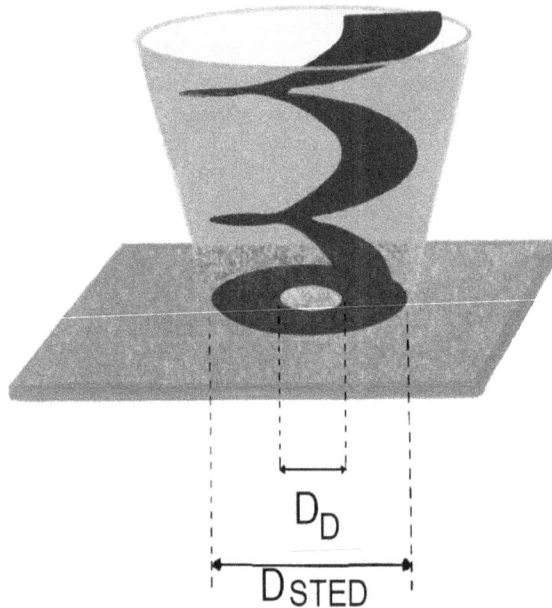

Figure 8.16. STED working principle. Picture credit: SumiB.

8.4.3 Lightsheet microscopy

In LSCM, an image of the sample was built point by point. The other commonly used imaging technique is that of widefield fluorescence microscopy (WFM) in which a large area is imaged, in one shot, onto a charge coupled device (CCD) or CMOS camera. The problem is that unlike the LSCM technique, out-of-focus light also reaches the camera, reducing the image contrast. Another issue that affects both WFM and LSCM is that the excitation light travels through the sample, causing fluorescence all along its path. Fluorophores in regions that are not being imaged (e.g. in planes away from the focal plane) start to fluoresce. This means they may not be available for fluorescence when required. Light sheet fluorescence microscopy (LSFM) [27] solves this problem by illuminating the sample from one side, while collecting the fluorescence from an orthogonal direction. A schematic of this is displayed in figure 8.17.

Because the fluorescence is collected at right angles to the direction of travel of the excitation light, extra optics are not needed to separate the two wavelengths, although filters may still be used to separate the reflected and fluorescent light. Since only a thin sheet of the sample is illuminated, no fluorophores from outside this region are illuminated, which provides two advantages. First, the fluorophores will suffer from less photobleaching and, second, there is no out-of-plane light that will reduce the image contrast. The aperture in LSCM was introduced to cut out this light, but that meant that scanning was required. With LSFM, an image of one entire plane of the sample can be acquired simultaneously, similar to the WFM technique, without having to worry about the out-of-focus light. The light sheet itself

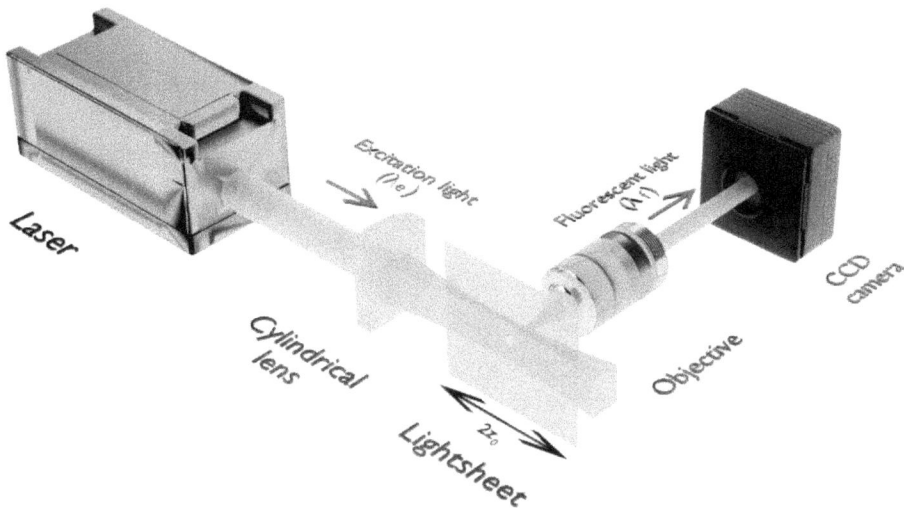

Figure 8.17. LSM working principle. Picture credit: JG.

is formed either by using a cylindrical lens, which is shown in figure 8.17 or by scanning a focused beam at the focal plane of the detection optics. In the early versions of LSFM, a Gaussian beam was used as the excitation beam and the FoV was determined by its Rayleigh range.

By now, it should be obvious that creating the light sheet with a Bessel or Airy beam instead of a Gaussian beam will have advantages. Both Bessel [28] and Airy [22] beams have been used in lightsheet microscopy. Their use not only helped increase the DOF or size of the light sheet, but also reduced the effect of scattering. Modifications of the beams have also been proposed and used that have further advantages [29].

References

[1] Durnin J, Miceli J J and Eberly J H 1987 Diffraction-free beams *Phys. Rev. Lett.* **58** 1499
[2] Hall D G 1996 Vector-beam solutions of Maxwell's wave equation *Opt. Lett.* **21** 9–11
[3] Schwarz U T, Bandres M A and Gutiérrez-Vega J C 2004 Observation of Ince–Gaussian modes in stable resonators *Opt. Lett.* **29** 1870–2
[4] Gutiérrez-Vega J C, Iturbe-Castillo M D and Chávez-Cerda S 2000 Alternative formulation for invariant optical fields: Mathieu beams *Opt. Lett.* **25** 1493–5
[5] Khonina S N, Kazanskiy N L, Karpeev S V and Butt M A 2020 Bessel beam: significance and applications—a progressive review *Micromachines* **11** 997
[6] McGloin D and Dholakia K 2005 Bessel beams: diffraction in a new light *Cont. Phys.* **46** 15–28
[7] Arlt J, Garcés-Chávez V, Sibbett W and Dholakia K 2001 Optical micromanipulation using a Bessel light beam *Opt. Commun.* **197** 239–45
[8] Shen Y, Pidishety S, Nape I and Dudley A 2022 Self-healing of structured light: a review *J. Opt.* **24** 103001

[9] Fahrbach F, Simon P and Rohrbach A 2010 Microscopy with self-reconstructing beams *Nat. Photon.* **4** 780–5

[10] Dharmavarapu R, Bhattacharya S and Juodkazis S 2018 Diffractive optics for axial intensity shaping of Bessel beams *J. Opt.* **20** 085606

[11] Paufler W, Böning B and Fritzsche St 2019 High harmonic generation with Laguerre–Gaussian beams *J. Opt.* **21** 094001

[12] Doster T and Watnik A T 2016 Laguerre–Gauss and Bessel–Gauss beams propagation through turbulence: analysis of channel efficiency *Appl. Opt.* **55** 10239–46

[13] Arfken G B and Weber H J 2005 *Mathematical Methods for Physicists* 6th edn (Amsterdam: Elsevier)

[14] Allen L, Babiker M, Lai W K and Lembessis V E 1996 Atom dynamics in multiple Laguerre–Gaussian beams *Phys. Rev.* A **54** 4259–70

[15] Shvedov V G, Desyatnikov A S, Rode A V, Izdebskaya Y V, Krolikowski W Z and Kivshar Y S 2010 Optical vortex beams for trapping and transport of particles in air *Appl. Phys.* A **100** 327–31

[16] Siviloglou G A, Broky J, Dogariu A and Christodoulides D N 2008 Ballistic dynamics of Airy beams *Opt. Lett.* **33** 207–9

[17] Berry M V and Balazs N L 1979 Nonspreading wave packets *Am. J. Phys.* **47** 264–7

[18] Efremidis N K, Chen Z, Segev M and Christodoulides D N 2019 Airy beams and accelerating waves: an overview of recent advances *Optica* **6** 686–701

[19] Siviloglou G A and Christodoulides D N 2007 Accelerating finite energy Airy beams *Opt. Lett.* **32** 979–81

[20] Broky J, Siviloglou G A, Dogariu A and Christodoulides D N 2007 Observation of accelerating Airy beams *Frontiers in Optics 2007/Laser Science 23/Organic Materials and Devices for Displays and Energy Conversion* (Washington, DC: Optica Publishing) p PDP_B3

[21] Introduction to the Airy functions *Wolfram MathWorld* https://mathworld.wolfram.com/AiryFunctions.html (Accessed: 17 June 2023)

[22] Vettenburg T, Dalgarno H I, Nylk J, Coll-Lladó C, Ferrier D E, Čižmár T, Gunn-Moore F J and Dholakia K 2014 Light-sheet microscopy using an Airy beam *Nat. Meth.* **11** 541–4

[23] Dharmavarapu R 2019 Improved techniques for complex light generation using diffractive and meta-surfaces *PhD Thesis* Indian Institute of Technology Madras

[24] Mehmood M Q *et al* 2016 Visible-frequency metasurface for structuring and spatially multiplexing optical vortices *Adv. Mat.* **28** 2533–9

[25] Ashkin A, Dziedzic J M, Bjorkholm J E and Chu S 1986 Observation of a single-beam gradient force optical trap for dielectric particles *Opt. Lett.* **11** 288–90

[26] Ayala A Y, Arzola V A and Volke-Sepúlveda K 2016 Comparative study of optical levitation traps: focused Bessel beam versus Gaussian beams *J. Opt. Soc. Am.* B **33** 1060–7

[27] Olarte O E, Andilla J, Gualda E J and Loza-Alvarez P 2018 Light-sheet microscopy: a tutorial *Adv. Opt. Photon.* **10** 111–79

[28] Olarte O E *et al* 2012 Image formation by linear and nonlinear digital scanned light-sheet fluorescence microscopy with Gaussian and Bessel beam profiles *Biomed. Opt. Exp.* **3** 1492–505

[29] George J G, Dholakia K and Bhattacharya S 2023 Generation of Bessel-like beams with reduced sidelobes for enhanced light-sheet microscopy *Opt. Cont.* **2** 1649–60

Chapter 9

Everyday optical systems and beyond

Through the course of this book we have looked at different aspects of optics. From chapter 6 onwards we also started discussing applications of light and optics. Clearly, there are a large number of tools, instruments, and diagnostic and measurement techniques that use light as an essential part of their systems. In this chapter, we explore a very small selection of applications that highlight the diverse and multi-faceted nature of light.

9.1 Barcode readers

Barcodes can be seen on almost every product we buy today. They are useful for keeping track of inventory. The black and white lines that are seen on a barcode represent the digits 0–9. An example is shown in figure 9.1. The light being used to interrogate the barcode is also visible.

Each digit is represented by seven equal-sized vertical blocks. Some of the blocks are white, the rest are black. There is a unique combination for each digit. For example, the number one is represented by two white stripes, followed by two black ones and then again two white stripes, and finally a single black stripe. The combination of white and black stripes have been chosen such that even if it is read upside-down, the same unique number is obtained. Barcode scanners have to be

Figure 9.1. An example of a barcode, typically seen on a product.

able to read the white and black lines on products and transfer that information to a computer. The basic principle is that light from a light emitting diode (LED) is incident on the barcode, and the reflected light is captured by a photodetector. White stripes reflect more light than black ones. A circuit captures this information and it is converted into 1s and 0s, which in turn gets converted into the digits that each combination represents. What complicates this rather simple device is that often the product with the barcode is moving while it is being scanned. This results in a blurred grey image rather than a clear set of black and white stripes. As with many other systems that use optics, computer algorithms are used to convert these unclear patterns into accurate barcodes.

9.2 Finger print sensors

Another very commonly used device is fingerprint sensors. They are used as a security measure. For example, you may need to use your fingerprint to access your cell phone or gain access to your workplace. Fingerprints are also captured for biometric purposes. That is one of the ways that physical characteristics can be used to identify individuals. Fingerprints can be captured using different techniques. Here, we discuss an optical technique that uses the principle of frustrated total internal reflection (FTIR) to create a high-contrast image of a fingerprint. The optics of such a sensor is shown in figure 9.2.

Light enters a glass prism at such an angle that it totally internally reflects the upper flat surface. However, when a person places their hand, in particular their fingers on the flat glass surface, the condition for FTIR is invalidated wherever the ridges (or raised portions of skin) touch the glass. That is to say, the refractive index of the prism n_g is not greater than that of the finger ridge, and light leaks out of the prism at these sites. Hence, the word *frustrated* in the description of this phenomenon. Between the ridges, total internal reflection continues to take place. Since light reflects better from the valleys between the ridges, a sharp contrast image of the fingerprint is created.

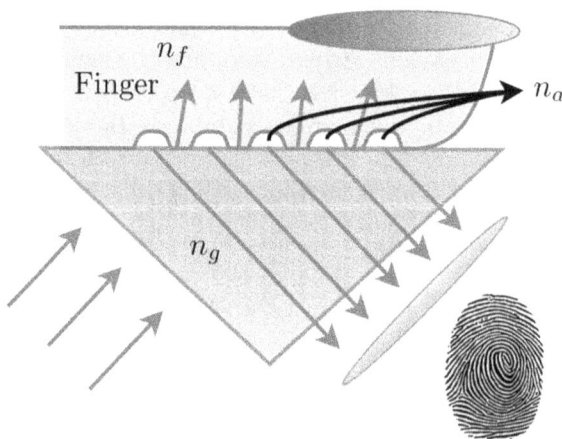

Figure 9.2. The optical system used to capture a person's fingerprint.

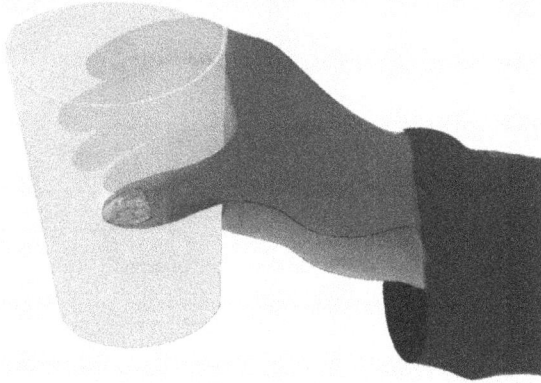

Figure 9.3. Observing ones fingerprint, with the help of a glass of cold water. Picture credit: SumiB.

Try this at home
   Take a transparent glass with no design on it and fill it with cold water. Hold it as shown in figure 9.3. You should be able to see your fingerprint clearly through the water, when looking from the top.

9.3 Pulse oximeters

Due to the COVID-19 pandemic, most people are familiar with this small but very important device. It is usually clipped onto a finger and is used to non-invasively measure the level of oxygen saturation of the blood. It should be noted that oxygen travels through blood by means of a protein called haemoglobin. Oxygen saturation indicates the percentage of haemoglobin that is carrying oxygen. 100% saturation would mean all of the available haemoglobin is carrying oxygen. A schematic of the device is shown in figure 9.4.

   The device uses two LEDs of different wavelengths. As light travels through the finger to the PD, on the other side, absorption occurs obeying Beer–Lambert's law:

$$I = I_0 \exp(-\alpha x). \tag{9.1}$$

I_0 is the starting intensity which reduces to I, after the beam has travelled a distance x in a medium with absorption coefficient α. The amount of absorption depends on the thickness of the finger, as well as the concentration of haemoglobin. Therefore, two wavelengths are used, as the absorption of these wavelengths is very different for blood saturated with oxygen compared to blood lacking oxygen. The wavelengths, of 660 nm and 940 nm, are chosen based on the fact that oxygenated haemoglobin absorbs more infrared (IR) light than red light, whereas the opposite is

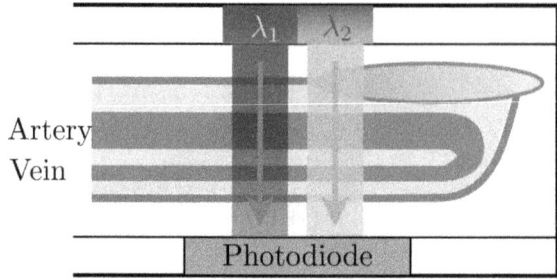

Figure 9.4. Working principle of a pulse oximeter.

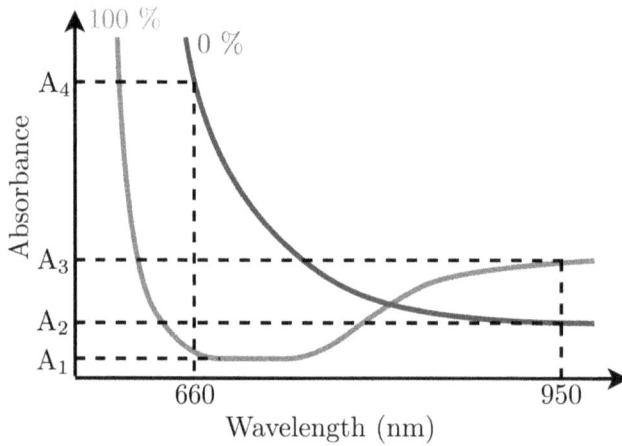

Figure 9.5. Graph showing change in absorbance versus wavelength for the cases of 0% and 100% saturation.

true for deoxygenated haemoglobin. Since everyone's finger has a different thickness, a pulse oximeter arrives at the oxygen saturation (SpO2) by comparing the difference in absorption between the red and infrared wavelengths. The kind of data that would be captured when calibrating such an instrument is shown in figure 9.5.

The ratios A_3/A_4 and A_1/A_2 depict the difference for the cases of 0% and 100% saturation, respectively. As these ratios vary from person to person, the pulse oximeter is pre-calibrated. The instrument will contain a look-up table from which the SpO2 value is picked based on the absorption ratios that are measured. The look-up table data is obtained from many different simultaneous measurements of light absorption and oxygen saturation carried out using alternate techniques. It is interesting to note that as the calibration depends on the measurement of the absorption of light, it has been found that there could be a certain racial bias to some systems [1]. The bias arises as the look-up table was generated from data collected from lighter-skinned individuals. This emphasises the need for diversity in data collection, particularly when the information is required for medical instrumentation.

9.4 Interferometry based measurements

Measurement is about recording parameters of an object or system of interest. By quantifying these parameters, one can then keep track of the system, as the parameters change and even modify the system based on the inputs received. When using light, the parameters that could be changing are intensity, phase, polarisation or even the wavelength. All of these are used in different systems. For example, polarisation sensors are used in industries as diverse as electronics (for detecting the presence of scratches on surfaces) to traffic regulation. Some very impressive images of the differences that can be observed by taking note of the polarisation of light can be seen on the Sony Polarsens website [2]. However, the most commonly used parameter, when using light as part of any measurement, is intensity. While this by itself is very useful in a large number of situations (any process that absorbs light will affect the intensity), it is also used to measure change in phase. A change in phase will not affect intensity directly. This is where interferometry comes into the picture. One beam of light undergoes the change in phase while interacting with the object under measurement and interference with a reference beam results in an intensity change. This indirect measurement is required, as present-day light sensors, such as photodetectors and cameras, measure intensity but not phase. These ideas were already discussed in chapter 6, and the use of interference in testing was dealt with in detail there. Here, we look at a few other applications.

9.4.1 Optical coherence tomography

Optical coherence tomography (OCT) is a non-invasive, cross-sectional imaging technique with micrometre resolution. It is used extensively in medical imaging and developmental biology to image tissue. Typical systems work in the infrared region (800–1600 nm). OCT offers better resolution than ultrasound, albeit at shallower depths.

3D images are assembled from point-by-point axial images acquired by scanning the incident beam across the sample. Each axial image is termed an 'A' scan. A compilation of a row of 'A' scans is called a 'B' scan, which gives a 2D image of one plane of the sample. Finally, by stacking B-scans together, a full 3D image can be obtained. These ideas are demonstrated in figure 9.6.

The heart of an OCT system, as with many other optical systems, is a Michelson interferometer. A schematic of the experimental set-up used for what is known as time domain optical coherence tomography (TDOCT) is shown in figure 9.7.

9.4.2 Time domain OCT

As with any interferometer, at least two beams of light are required. In the case of TDOCT, one beam reflects off the sample, picking up the desired information and the reference beam reflects off a mirror, which is moveable. The reference beam intensity at the detector is denoted as I_R. As shown in figure 9.7, initially, the beam is incident at one point of the sample. The sample beam consists of light reflected from all the scatterers (I_{S1}, I_{S2}, I_{S3}, etc) along the axis. The light that returns to the detector

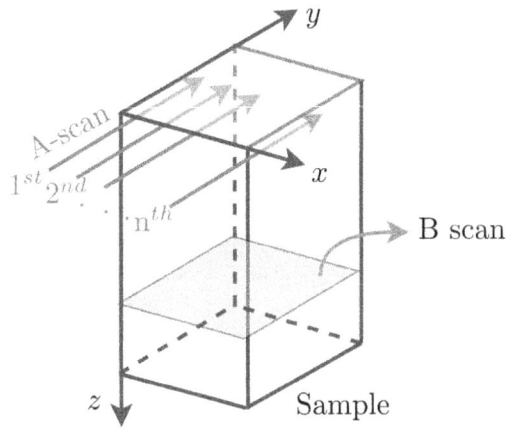

Figure 9.6. Schematic showing an A-scan, B-scan and full 3D image of a sample.

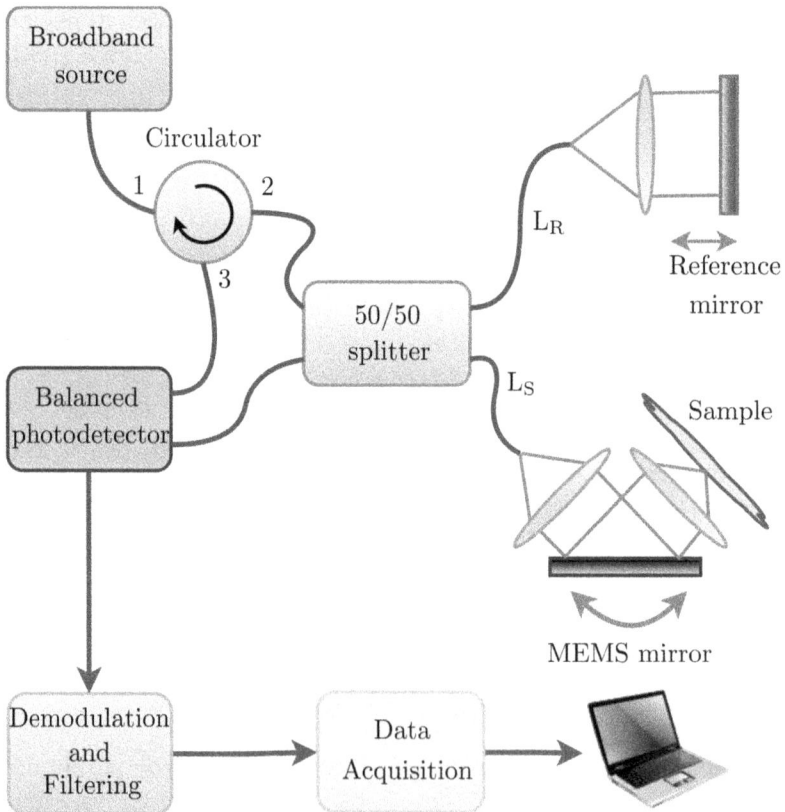

Figure 9.7. Schematic of Michelson interferometer based time domain OCT system.

from that point can be thought of as a set of superposed beams. In order to acquire information about each scatterer individually, a low coherence or broadband source is used, instead of a laser. Because of the low coherence, only the scatterer whose optical path length matches that of the reference mirror will contribute to the interference signal. This idea is known as coherence gating. In order to get the information from other scatterers along the path, the reference mirror is moved along the axis. Each position provides information about a different depth (or scatterer) of the sample. In this way, axial information for one (x, y) position on the sample is acquired. The interference equation is similar to equation (6.6) and consists of a dc background plus the interference term. The interference intensity is given in terms of the reference mirror position (z_R) as

$$
\begin{aligned}
I(z_R) = &\left[\left[I_R + I_{S1} + I_{S2} + I_{S3} + \cdots\right]\right] \text{DC terms} \\
&+ 2\left[\sum_{n=1}^{N} \sqrt{I_R I_{Sn}} \exp(-(z_R - z_{Sn})^2 \Delta k^2)\cos(2k_0(z_R - z_{Sn}))\right] \text{interference term,}
\end{aligned}
\tag{9.2}
$$

where k_0 is the central wavenumber of the spectrum.

It should be noted that in each case the intensity could be specified in terms of $S_0 R_i$, where S_0 is the spectral power integrated over all the wavelengths of the source and R is the reflectivity of either the reference mirror or the scatterers in the sample. The $z_R - z_{Sn}$ term in the argument of the cosine is the parameter we are interested in. The reference mirror location tells us where a scatterer is in the sample and the $\sqrt{I_R I_{Sn}}$ term indicates the strength of that scatterer.

The key difference between this equation and equation (6.6) is the $\exp\left[-(z_R - z_{Sn})^2 \Delta k^2\right]$ term, usually denoted by $\gamma(z)$, which is the coherence function of the source. Clearly, the larger this difference is, the smaller the value of the exponential and this is the basis of TDOCT. If the reference mirror is not path-length matched to the scatterer, that scatterer contributes weakly if at all to the interference term. (It will always contribute to the dc term.) More details of this are available in the chapter on the theory of optical coherence tomography by Izatt and Choma [3]. In equation (6.6), the source was assumed to be perfectly coherent and therefore, $\gamma(z) = 1$.

The limitation of TDOCT is that, in addition to the lateral scanning, the reference mirror is also moved to obtain information along the axis. This makes the process slow. An adaptation that overcomes this is called Fourier domain optical coherence tomography (FDOCT).

9.4.3 Fourier domain OCT

In TDOCT information is acquired by varying z_R in equation (9.2). This variation is used to create the image. In FDOCT, on the other hand, the reference mirror is not moved. The change in the equation is brought about by recognising that the wavevector k appears in the interference term of equation (9.2). By replacing the photodetector in the experiment with a spectrometer, the wavelength information (present but lost in TDOCT) is acquired. Why is this important? To understand this, let us look at the intensity versus wavelength for two beams assuming the experiment used a broadband source and a spectrometer as the detector. In addition, if the sample was a single reflector located at z_1 and pathlength matched with the reference mirror, the equation to represent the interference would be

$$I(k) = I_R + I_{z_1} + 2\sqrt{I_R I_{z_1}}\cos(2k(z_R - z_1)). \qquad (9.3)$$

The resulting intensity versus k plot would appear something like the one shown in figure 9.8(a).

This plot would change for each disturbance located along the same axial length of the sample but at a different depth of the sample. For example, if the scatterer was located at z_2 or z_3 instead of z_1, the image would be like figures 9.8(b) and (c), respectively. Careful observation of these plots reveals that the frequency of the fringes is dependent on the depth of the scatterer being imaged. This is the basis of Fourier domain OCT. In an actual sample, the information from all scatterers will be received simultaneously, as shown in figure 9.9.

A Fourier transform of this signal will provide us with the frequencies present. This information can be used to locate each scatterer, the amplitude of which indicates the strength of each scatterer. This is shown in figure 9.10, where the sample is assumed to consist of the three layers that had been studied independently. In this graph, the FT frequencies have been converted to distances using the equation

$$z = \frac{\pi}{\Delta k}\left[\frac{-N}{2} : \frac{N}{2}\right], \qquad (9.4)$$

where N is the number of pixels.

The current [3] on each pixel of the spectrometer of the FDOCT experiment (caused by the incident intensity) is given in terms of the wavenumber k as

$$i(k) = \frac{\rho}{4}\left[S(k)\left[R_R + R_{S1} + R_{S2} + R_{S3} + \cdots\right]\right] \text{ DC terms}$$

$$+ \frac{\rho}{2}\left[S(k)\sum_{n=1}^{N}\sqrt{R_R R_{Sn}}\cos(2k(z_R - z_{Sn}))\right] \text{ cross correlation terms} \qquad (9.5)$$

$$+ \frac{\rho}{4}\left[S(k)\sum_{m\neq n=1}^{N}\sqrt{R_{Sm} R_{Sn}}\cos(2k(z_{Sm} - z_{Sn}))\right] \text{ Auto correlation terms}$$

ρ is the detector responsivity in amp/W, $S(k)$ is the power spectral distribution of the light source, R_R and R_{Si} are the mirror and ith scatterer reflectivities, respectively.

FDOCT enables the fast capture of 3D volumetric images. In general, however, it works over shorter distances than TDOCT. Clever engineering and algorithms are used to overcome issues such as sensitivity roll-off, etc. Commercial OCT systems are available in all major hospitals across the world today and are used extensively in ophthalmology and cardiology.

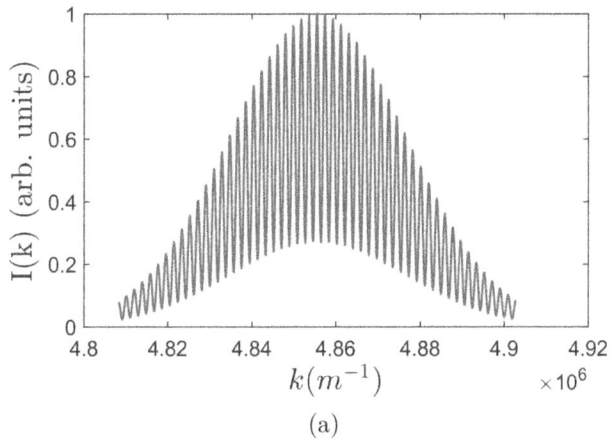

(a)

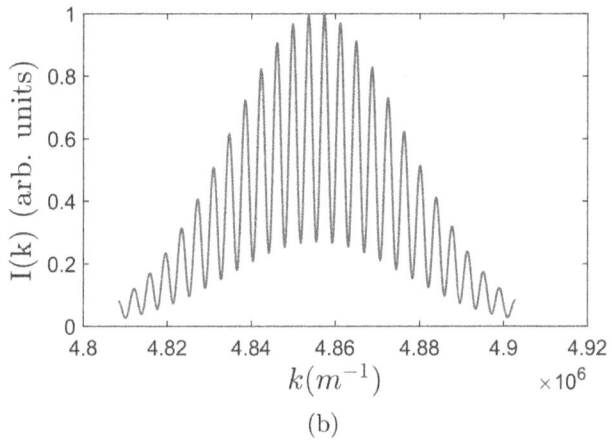

(b)

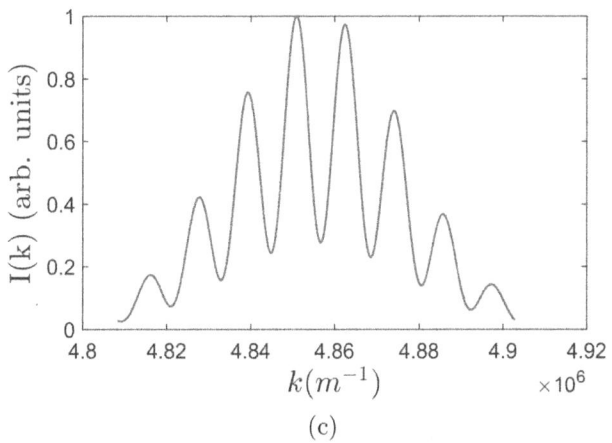

(c)

Figure 9.8. Graphs show the intensity versus k plots for an FDOCT system. The only difference in each case is that the sample is at a different depth in each case: (a) at z_1, (b) at z_2, and (c) at z_3, where $z_1 > z_2 > z_3$. OCT data from NP.

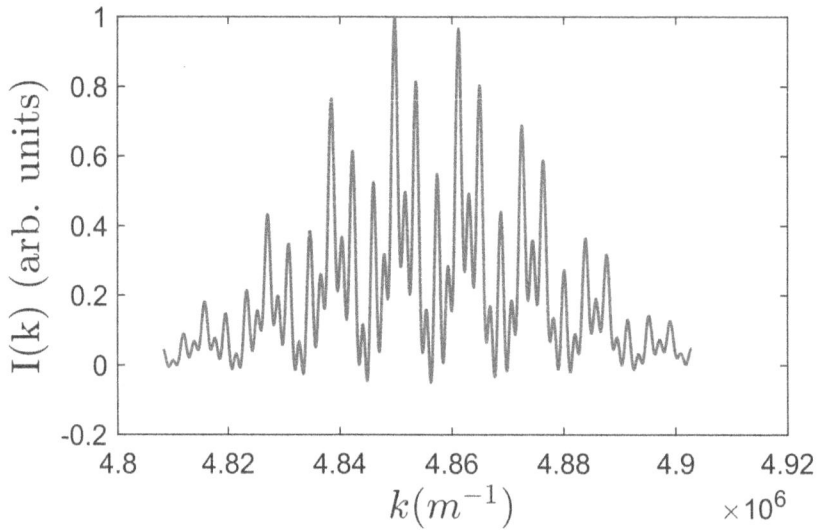

Figure 9.9. Interference pattern obtained when the sample simultaneously contains all the layers shown in figure 9.8. OCT data from NP.

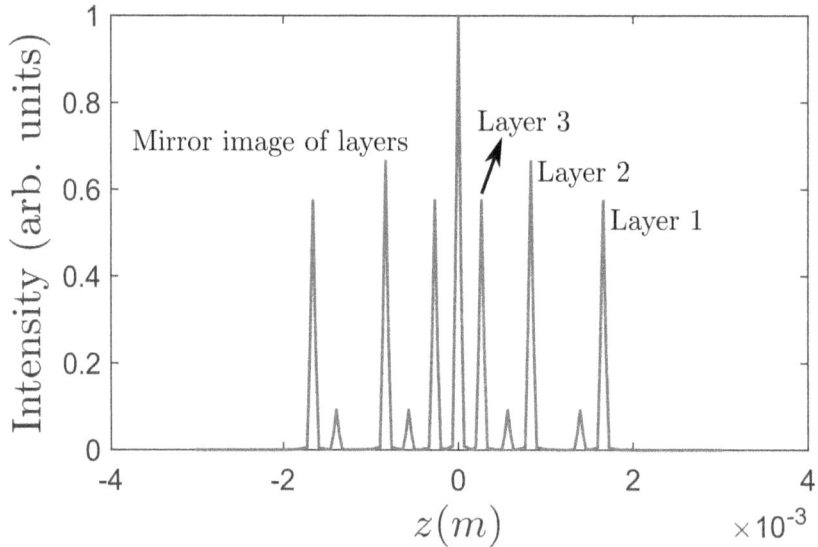

Figure 9.10. FT of the interference pattern obtained when the sample simultaneously contains all the layers shown in figure 9.8. OCT data from NP.

9.4.4 Fourier transform infrared spectrometry

Many molecules have characteristic absorption peaks in the IR region. This absorption pattern acts like a unique fingerprint of the molecule. In other words, capturing the absorption peaks of a sample with respect to wavelength allows scientists to identify the molecules present. A standard instrument used for this purpose is a spectrometer. There are many different types of spectrometers. Commonly used in the IR region are instruments based on the principle of Fourier transform spectrometry (FTS). The instruments are so called because a Fourier transform of the output (obtained as an intensity signal versus time) can be used to find out the absorption peaks versus wavelength. The system has several advantages over other spectrometers, the most important one being that information about all the absorption wavelengths is obtained in one shot, without have to mechanically rotate a prism or grating. The improved signal to noise ratio (SNR) obtained by acquiring this information in one measurement is known as the Fellgett advantage [4].

The core of such a system is once again a Michelson interferometer. In this case, however, we are interested in analysing the changes in amplitude in the wavelengths present in the source after it passes through a sample. For this purpose, a conventional Michelson interferometer, as shown in figure 6.4, can be used, the only difference would be that the light from the source would travel through the sample of interest before entering the interferometer. The FT in the name of this technique suggests that the desired information is going to be extracted using frequencies. This is indeed the case and is achieved by actuating one mirror of the interferometer with a periodic signal of frequency f_m. The mirror is made to move in an out-of-plane direction. If one assumes that the mirror moves with a constant velocity v_m, the optical path difference between the two beams of the interferometer will vary with respect to time as $2v_m t$ and the corresponding phase difference δ will be

$$\delta(t) = \frac{2\pi}{\lambda} 2v_m t. \tag{9.6}$$

The factor of 2 comes about because the change in optical path length will be twice the displacement of the mirror. Just as was seen in (6.6), the intensity at the photodetector will be

$$I(t) = I_0[1 + V\cos(\delta(t))]. \tag{9.7}$$

It should be noted that this equation will be valid for each wavelength present in the source. The sinusoidal component of this equation is the term that carries the desired information. The argument of the cosine can be written in terms of wavenumber k

$$\delta(t) = 2kv_m t. \tag{9.8}$$

Given the assumption of constant velocity movement of the mirror, the signal at the detector will be found to vary sinusoidally with a different frequency f_{FFT} for each wavelength present in the source. This is because the interference pattern will be maximum whenever the OPD is an integral multiple of each of the wavelengths present in the source. It is known that a cosine signal $S(t)$ with amplitude S_0 and frequency f_{FFT} can be written as

$$S(t) = S_0 \cos(2\pi f_{FFT} t) \tag{9.9}$$

Therefore, comparing equations (9.8) and (9.9), for any one wavelength, we see that

$$2\pi f_{\text{FFT}}\, t = 2kv_{\text{m}} t \qquad (9.10)$$

or

$$f_{\text{FFT}} = \frac{2v_{\text{m}}}{\lambda}. \qquad (9.11)$$

As the actual source consists of a range of wavelengths, the measured intensity signal consists of the superposition of all the individual interferograms. A FT of this signal will yield a number of peaks corresponding to the main wavelengths present. This signal contains frequency and amplitude information. The former is used to identify the sample with which the light had interacted. On the other hand, the amplitudes of the peaks indicate the strength or concentration of molecules being detected. In essence, the moving mirror helps down-convert the frequencies of the light beam into frequencies measurable by present-day photodetectors. As seen in equation (9.11), the values of the FT frequencies will be determined by the wavelengths, as well as the mirror velocity.

This section presents a rather simplified version of FTS. In reality, the amplitude of the FT peaks will depend on other factors, such as efficiency of components such as the beamsplitter, the detector response, etc. Also, the assumption of the mirror moving at constant velocity is not valid. A more rigorous discussion can be found in [5].

References

[1] Sjoding M W, Dickson R P, Iwashyna T J, Gay S E and Valley T S 2020 Racial bias in pulse oximetry measurement *New Engl. J. Med.* **383** 2477–8
[2] Sony Polarization Image Sensor Technology Polarsens *Sony* http://www.sony-semicon.com/en/technology/industry/polarsens.html (Accessed: 2 February 2024)
[3] Drexler W and Fujimoto J G 2008 Optical coherence tomography: technology and applications *Biomedical Engineering* (Berlin: Springer)
[4] Fellgett P 1951 *PhD Thesis* University of Cambridge
[5] Griffiths P R, De Haseth J A and Winefordner J D 2007 Fourier transform infrared spectrometry *Chemical Analysis: A Series of Monographs on Analytical Chemistry and Its Applications* (New York: Wiley)

Appendix A

Using Zemax or OSLO to achieve optical design goals

A.1 Designing a lens with a specific focal length

In this example, the design tools are being used to create a plano-convex lens with a focal length of 100 mm (table A.1).

First, the correct operand has to be selected ('EFFL' in Zemax or 'EFL' in OSLO) that tells the software that we want to set the effective focal length of the system to a particular value. Then appropriate variables have to chosen. These are parameters that can be varied (without changing our application requirements) that will force the the focal length of a lens to have a particular value.

In OSLO, the 'PU' operand controls the angle of the axial ray. This could be used instead of 'EFL' to change the focal length. Assuming the light incident on the lens is a well-collimated beam, and that a single thin lens is being designed, the 'PU' operand can be used to force the lens to have a certain focal length. This is because in that case, the angle of the axial ray will be determined by the ratio of the radius of the beam to the focal length.

A.2 Drawing the principle planes in an optical system

A.2.1 In Zemax

Once the optical system is designed, the principal planes can be made visible in Zemax in the following way. First, the cardinal point data report is to be generated. Let us look at the report generated for problem 6 in chapter 3, which is given in figure A.1.

This report gives us information about the location of the principal planes with respect to other planes. We will use this information to draw a virtual plane in the layout. To do so, two rows have to be added to the lens layout for each principal plane. They will be added between the last row (image plane) and the preceding row. Let us assume these were rows 5 and 6, respectively. Inserting a row here creates a new sixth row. In this row, a solve is used to enter details in the thickness cell. To display the first principal plane (PP) the position solve (given as 'T' next to the

Table A.1. Optimisation techniques in Zemax and OSLO.

Zemax	OSLO
Click on 'Optimize' from the main menu. Open the merit function editor window: • Select operand type as 'EFFL'. • Enter weight as 1.	Select 'Generate Error Function' and then: • Select 'Aberration Operands'. A list of default operands will appear. • The 'EFL' or the 'PU' operand can be chosen. • Make the weight of the chosen operand 1.
• Enter operand target as 100.	• In OSLO, the operand definition is the error function that is to be made 0. The EFL operand is denoted by OCM21. Therefore, its definition to obtain a focal length of 100 has to be OCM21-100.
Make the radius of curvature of the first surface a variable. • Click 'Optimize'. • Click on 'Start'. • Once optimisation is complete, note that the merit function is reduced to zero. • Click on 'Exit'.	• Close the spreadsheet, and click the 'Open' button in the text window. Selected operands with their current values will be displayed. • Click the 'Ite' button in the text window to optimize.
The value of the effective focal length is 100, indicating that the variables selected were appropriate for solving the problem.	

```
Object space positions are measured with respect to surface 2.
Image space positions are measured with respect to surface 10.
The index in both the object space and image space is considered.

                              Object Space      Image Space
Focal Length          :      -150.000000       150.000000
Focal Planes          :      -171.981051        -0.000000
Principal Planes      :       -21.981051      -150.000000
Anti-Principal Planes :      -321.981051       150.000000
Nodal Planes          :       -21.981051      -150.000000
Anti-Nodal Planes     :      -321.981051       150.000000
```

Figure A.1. Cardinal point data.

value of thickness) is chosen. For this particular example, a distance of −21.981 from surface 2, obtained from figure A.1, is entered. If nothing else were done at this point, the ray diagram would look quite odd. The next row added (which will be row 7), therefore, needs to undo the effect surface 6 had on the rays. The thickness for row 7 is arrived at using a pickup solve (given as 'P'), wherein the thickness of surface 6 with a scale factor of −1 is selected. The addition of these two rows allows us to use the tool to draw a physical surface at the location of the first principal plane. Since surfaces 6 and 7 are dummy surfaces that do not play any optical role in the design, one must open the surface properties menu and select both 'Hide Rays To This Surface' and 'Skip Rays To This Surface', as shown in figure A.2.

The same exercise is repeated to include the second PP. In figure A.1, the distance of this plane (−150 mm) is given with reference to a surface (10) that appears *after* it. Since distances have to be given with respect to an earlier surface, this number cannot be used directly. Instead, we take the sum of the distance of PP2 given in the cardinal point report and the thickness of the last real surface (namely, surface 5 in this case), which is −150 + 117.135, yielding a distance of −32.865 from surface 5. The details of the lens entry are visible in figure A.3.

The final lens layout with principal planes displayed is shown in figure A.4.

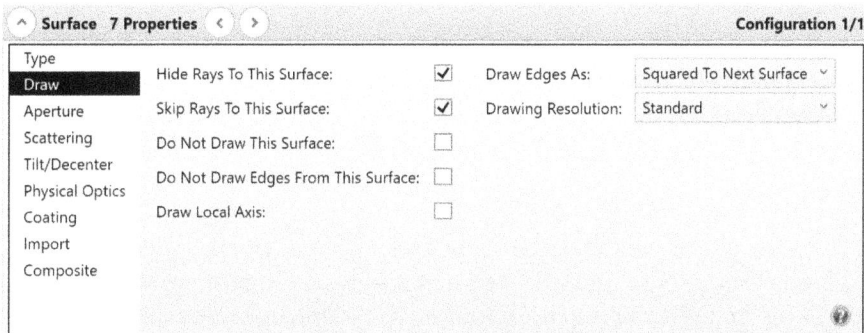

Figure A.2. Surface properties menu in Zemax.

	Surface Type	Comment	Radius	Thickness	Material	Coating	Clear Semi-Dia
0	OBJECT Standard ▾		Infinity	Infinity			Infinity
1	Standard ▾		Infinity	40.000			10.000 U
2	STOP Standard ▾		89.672 V	5.000	N-BK7		5.025
3	Standard ▾		-45.000	10.000			5.456
4	Standard ▾		-70.000	4.000	N-BK7		6.275
5	Standard ▾		91.645 V	117.135 M			6.656
6	Standard ▾		Infinity	-158.117 T			27.471
7	Standard ▾		Infinity	158.117 P			10.982
8	Standard ▾		Infinity	-150.000 T			27.471
9	Standard	Thickness solve on surface 8					9.008
10	IMAGE Standard						27.471

Solve Type: Position

From Surf: 5

Length: -32.865

Figure A.3. Details of lens data showing the extra rows needed to physically represent the principal planes. In row 8, the thickness option has been highlighted.

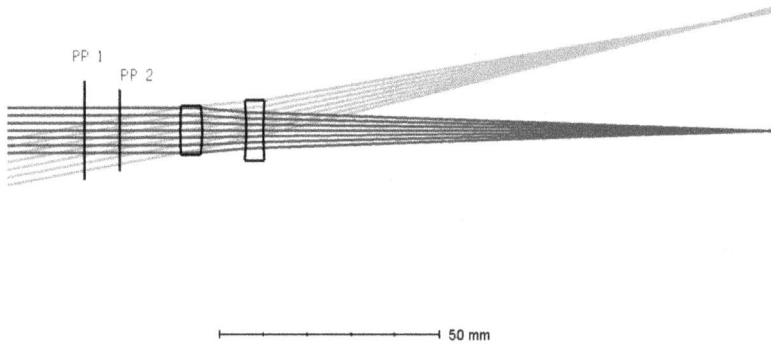

Figure A.4. Lens layout with principal planes displayed.

A.2.2 In OSLO

The location of the principal planes can be obtained from the 'Setup menu'. While dummy surfaces can be included at these positions, in versions 6.6.5, or lower, rays to these surfaces cannot be hidden, resulting in strange figures!

A.3 Optimising aberrations

Each software allows one to select a set of operands that will be minimised during an optimisation process. For example, you might want to minimise spherical aberration of a primarily on-axis system. The optimisation is done by setting certain parameters (such as the radius of curvature of a surface) to be variable and not fixed. It is often important whilst doing this to ensure the focal length of the system does not change, as the purpose is to minimise the aberrations of a certain system. Changing the focal length would most likely change the system itself.

A.3.1 In OSLO

Here, one would open the 'Operands Data Editor' as shown in figure A.5. As can be seen from the table, a number of operands can be selected simultaneously, along with a weight (of maximum value 1) that determines their importance in the calculation of the error function.

In both software, variables have to be selected that will be varied in an attempt to achieve the desired values of the operands. The idea behind both optimisation routines is similar. However, in the case of OSLO, the operand will be minimised to 0. Nothing needs to be done for the SA3 operand in OSLO, as in most cases, one wants that to be 0. However, if one was optimising the effective focal length (EFL) to 100 mm or just ensuring that it remains at this value, then the definition would be given as OCM21-100, so that once the optimisation is done, the final value of the operand will be 0.

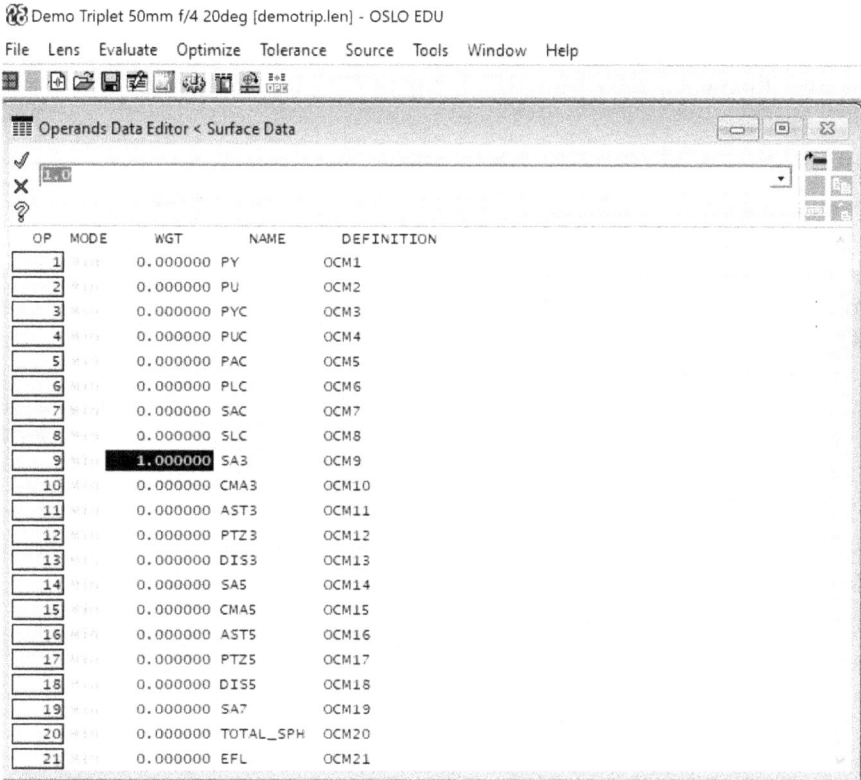

Figure A.5. 'Operands Data Editor' for defining the aberration operands in OSLO.

A.3.2 In Zemax

The equivalent to the OSLO 'Operands Data Editor' is the 'Merit Function Editor' as seen in figure A.6.

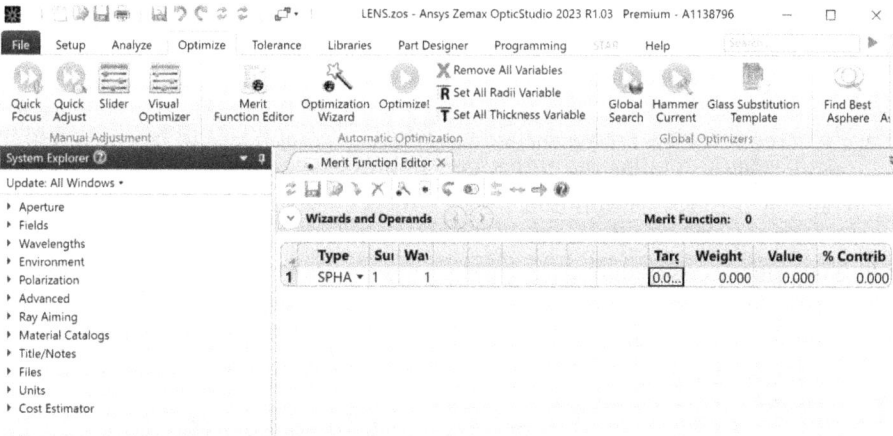

Figure A.6. 'Merit Function Editor' for defining the aberration operands in Zemax.

In Zemax, one needs to set the desired target value in the appropriate cell of the table. Alternatively, one can use the Optimisation Wizard to optimise for the RMS spot, rather than enter the operands directly in the Merit Function Editor.

A.4 Optimising chromatic aberration

Chromatic aberration can be corrected by the combined use of convex and concave lenses, as well as lenses of different glass types, i.e. glasses with high and low V-numbers. Its easy enough to set the radii of curvature of a lens as a variable and optimise an optical system based on that. However, a really powerful option of both OSLO and Zemax is that one can also use the software to help choose the best glass type to minimise chromatic aberration. The procedures for both software are given assuming that a doublet of two different materials has been entered in the spreadsheet. All basic lens parameters have been set and operands have been chosen such that the EFL of the system will not change during the chromatic aberration correction.

A.4.1 In Zemax

This is done using the following sequence of steps:
- First one selects the 'Libraries' option.
- Following this, select the 'Glass Substitution Template'. This allows one to define a template to add some constraints to one's choice of glass.
- A 'Substitute Glass Solve' is required to tell OpticStudio to exchange the initially selected glasses for any other glass that meets the substitute template's requirements.
- As with any optimisation, other variables such as the radii of curvature are set.
- Optimisation for glass substitution is carried out using the 'Hammer' optimisation command. As the name suggests this is not a refined operation and continues until the user chooses to end the operation.

A.4.2 In OSLO

Details of this process are available in the OSLO user guide [1].
- To make the glass type a variable in OSLO, it first needs to be converted to a 'model' glass. Under the 'GLASS' column of the surface data spreadsheet, click on the grey button and select 'Model...(M)'.
- Accept the default glass name, refractive index and V-number.
- Now, one needs to define the glass as a variable. Once again pressing the grey button in the glass column will show a list of variables. Select the variable 'RN/DN'. Although the V-number 'RN' is used to define the dispersion for the model glass, it is the normalised parameter 'DN' which is used as a variable.
- Click on the variables button at the top of the spreadsheet and set an upper limit of 1.0 for the value of DN.

- One can set the limits of RN to reasonable values, in terms of the cost or environmental properties of the glass.
- Finally, one sets the radii of curvature as variables and optimises the design.
- The exercise has created a fictitious glass that will reduce the chromatic aberration. One needs to select a real glass whose RN will closely match based on the DN value of the generated glass.
- By clicking on the grey button for the glass type that was being optimised, the 'FIX' option becomes available. This allows one to select a real glass type with a matching or closely matching V-number from the latest glass catalogs available in your edition of OSLO.

Reference

[1] Oslo Optics Reference *Lambdares.com* https://lambdaresfiles.com/wp-content/uploads/support/oslo/oslo_releases/OSLOOpticsReference.pdf (Accessed: 29 March 2023)

www.ingramcontent.com/pod-product-compliance
Lightning Source LLC
Chambersburg PA
CBHW080540220326
41599CB00032B/6327